Memoirs of a Cold Warrior

Memoirs of a Cold Warrior

The Struggle for Nuclear Parity

Lee C. Carpenter

Algora Publishing
New York

Library of Congress Cataloging-in-Publication Data —

Carpenter, Lee, 1925-
Memoirs of a cold warrior : the struggle for nuclear parity / Lee Carpenter.
p. cm.
Includes index.
ISBN 978-0-87586-702-1 (trade paper: alk. paper) — ISBN 978-0-87586-703-8 (hard cover: alk. paper) — ISBN 978-0-87586-704-5 (ebook) 1. Nuclear weapons—Government policy—United States—History. 2. Ballistic missiles—Design and construction. 3. Intercontinental ballistic missiles—United States. 4. Intercontinental ballistic missiles—Soviet Union 5. United States—Foreign relations—Soviet Union. 6. Soviet Union—Foreign relations—United States. 7. Carpenter, Lee, 1925- 8. Aerospace engineers—United States—Biography. I. Title.

UA23.C268 2009
355.02'17097309045—dc22

2009003053

Front Cover, Left: Lee C. Carpenter (author's collection);
Right: Intercontinental Ballistic Missile in Silo © Marvin Koner/CORBIS, ca. 1968

Printed in the United States

This work is dedicated to the courageous, proficient and diligent men and women in the military services of the United States of America.

Table of Contents

Preface 1

Chapter 1. Nike Air Defense 3
Siting 5
New York 7
Washington 7
San Francisco 8
Bell Telephone Laboratories 9

Chapter 2. IBM Military Programs 11
Hardware Development 11
B–70 Bombing-Navigation System 13
Demise of the B–70 17
Civil Defense and Shelters 20

Chapter 3. The Space Guidance Center and Nuclear-Powered Cruise Missiles 23
SLAM 24
Pluto Reactor 27
Radiation-Resistant Computer 27

Chapter 4. Strategic Bombers, Missiles & Armaments 29
B–52 Re-Instrumentation Study 30
Air Force Priorities 31
The Small Air-to-Surface Missile, or SRAM 32
Electro-Optical Sensors 32
Multipurpose Long Endurance Aircraft 33

IBM Diversification 34
Consequences of the B–52 Study 34
Boeing Company 34
Martin Marietta Company 34
New Business Plan 35
A New Strategic Bomber 36
Manned Aircraft Strategic Systems Group (MASSG) 37
Air Staff & SPO Support 40
SAWSEU II 41
Chapter 5. New Bomber Avionics 45
Shaping Public Opinion 45
AMSA Avionics Proposal 47
ASSB Program Growth 48
Experiment Five 49
Experiment Six: Simulation Laboratory 50
Tactical Avionics Programs 51

CHAPTER 6. A–7D/E AVIONICS PROGRAM 53
A–7D/E Avionics Proposals 53
Competition 54
Final Phase 55
Negotiations 56
On Notice 58
Open Door Policy 60
Survival 60
Last Interview 62

CHAPTER 7. BOMBER & AWACS ADVOCACY, AND MORE AVIONICS 65
The B–52 Program 65
Titan III Computer 66
AMSA Avionics Program 66
AWACS, or Airborne Warning and Control System 67
AMSA Program 69
B–52 Transfer 71
B–1 Advocacy 71
B–1 Liaison & Reviews 72
Management 74

CHAPTER 8. B–1 AVIONICS COMPETITIONS 77
B–1 Avionics Competition 78
End Game 78
Market Requirements 79
B–1 Advocacy 80
B–1 Avionics Finale 83
End Game 89

Chapter 9. Program Analyses & Selection 93
Space Shuttle 93
SECRAC 94
Higher Defense Costs 95
World Trends and Program Selectivity 96
1973 Israeli War 98

Chapter 10. Air Force Invitational Orders 101
B–1 Advocacy Study 101
Studies 103
World-Wide Military Command & Control System–II (WWMCCS-II) 105
Renewed Orders 105
US & USSR Strategic Forces Study 106
Counterforce Model 106

Chapter 11. Program Management & Congressional Liaison 113
B–1 Computer 114
Technology Transfer 116
MI 10-24 117
Defense Science Board 117
B–1 Program in Congress 118
Soviet Strategic Forces 119
US Strategic Forces 119
MX-CCC Red Team 121

Chapter 12. New Invitational Orders 123
Methodology 125
Threat Synthesis 126
GOSG 126
Preliminary Results 126
Target Structures 127
Force Effectiveness 127
Clerical Comparisons 128
Counterforce Calculations 128
End Game 130
Audits & Presentations 131
SAC 132
Follow On 134
Ambassador Watson 134
Reagan Transition 135
Drumming Up Support 135

CHAPTER 13. REAGAN TRANSITION TEAM 137
Threats 137
Liaison 138
USAF Aeronautical Systems Division 139
Protocols 139
Defense Science Board 140
USAF Headquarters 140
W. H. Taft, IV 141
OSD Panel 141
Townes Committee 141
DIA Liaison 142
Final DOD Initiatives 142
Civil Agencies 143

CHAPTER 15. SENATE TESTIMONY & HOUSE STUDY 145
Senator Tower 145
Senator Laxalt and Senator Garn 145
Testimony 146
Censors 147
Lt. Colonel Montoulli 147
Senator Tower Reacts 149
The House 150
Interior Committee Staff Study 151
End Game 151
Think Tanks 151
Thomas Watson, Jr. 152

CHAPTER 15. REAGAN'S DECISIONS, & SENATE TESTIMONY 153
President Reagan's Decisions 153
The Press and the Public React 154
Joint Chiefs 155
HASC 155
Senate Testimony 156
Hearings 156
Reactions 157
Boeing 158

CHAPTER 16. HOUSE TESTIMONY & ICBM BASING 161
HASC Testimony 162
Ron Mann 163
T.K. Jones, OSD 163
MX Basing 163
Dense Pack 164

CHAPTER 17. MILITARY, INDUSTRIAL & CONGRESSIONAL LIAISON 167
Army BMD 167
Hudson Institute 168
HASC 168
Boston Globe 170
Defense Week 170
Townes II 171
NSC 171

CHAPTER 18. IBM EXODUS 173
STARS Proposal Team 174

CHAPTER 19. THE PRESIDENT'S COMMISSION ON STRATEGIC FORCES & TESTIMONY TO THE HOUSE 177
HASC 179
To Harden or Not to Harden 180
Commission's Report 180
House III 181
Hearings 181
The Soviets 182
Testimony 182
B-1 Bombers 183

CHAPTER 20. US ARMS CONTROL AND DISARMAMENT AGENCY, ACDA 185
ASAT-Option 9 186
Verne Wattawa 186
SDI Tour 186
Negotiating with the Soviets 187
Supporting Groups 188
NORAD & DSB 188
Public Diplomacy 188
The official platform was that: 189

CHAPTER 21. NST ROUND III — GENEVA 191
Ambassador Kampelman 191
First Plenaries 192
Soviet Surprise 194
ACDA 195
Assignments 196

CHAPTER 22. NST ROUNDS IV–VI 199
Round IV 199
Contention 200
Disaster in Space 201

Soviet Initiative 201
First Interim 202
Washington Lunches 202
Transition Study 203
Round V 203
Summer Break 204
Round VI 205
End of Round 206

CHAPTER 23. NST ROUND VII & SPACE POLICY 209
Negotiations 210
New Milieu 210
Back-Stopping 211
Space Policy 211
Next 212

CHAPTER 24. NST ROUNDS IX & X 213
Analyses 214
Negotiations 214
End of Round 216
Round X 217
Stateside–1 217
Geneva 218
SDI 218
Acting Executive Secretary 219
Stateside–2 219
Geneva Again & Again 219
Stateside–3 219

CHAPTER 25. ONE LAST STRATEGIC STUDY 221

CHAPTER 26. STRATEGIC NUCLEAR PARITY 225
Today's Balance of Power 225
Milieu of the US Armed Services 227
National Problems 227

ACKNOWLEDGMENTS 229

Preface

As Russia re-asserts itself on the global stage, and now the Peoples Republic of China, too, a look back at the hard, cold facts of the Cold War may improve Americans' understanding of our relative strengths and weaknesses and the continuing vulnerability of our primacy in the world. An engineer who served on the frontlines of the struggle for military parity, I was party to and motivated many of the steps taken by US military, technical and industrial communities to assess, counter, and of course to seek to outperform the Soviet Union from 1952 to 1989. The US was the dominant power during the 1950s and 1960s, but the Soviets established overwhelming strategic military superiority in the 1970s and 1980s. The subsequent collapse of the Soviet Union, combined with successful arms negotiations, have established stability and essential strategic parity during the past twenty odd years.

The Cold War, which Americans consider began when the Soviet Union occupied Eastern Europe, blockaded Berlin, and supported North Korea during the "United Nation's Police Action," had clearly established the Soviet Union as a long-term strategic threat to the United States and our allies. As our most important national problem, helping to contain this threat was a vital endeavor and to it I devoted my professional lifetime.

The various interested parties who worked throughout the Cold War years on projects like Nike, the B-70, and other defense systems make up the "Military–Industrial Complex"; they provided the main cast of characters in this book.

After serving in the US Marine Corps in World War II and as a US Naval Technician during the Korean War, I developed a long professional career with the Bell Telephone Companies, IBM, the US Arms Control and Disarmament Agency, and the US Department of State, dedicated to enhancing the strategic defense posture of the United States. More important, I served as a voluntary defense consultant for over 30 years, advising the US Army, Air Force, Department of Defense, President Reagan's transition team, subcommittees of the Senate Arms Services and the House of Representatives Interior and Armed Services Committees, and President Reagan's Commission on Strategic Forces. For the US Arms Control and Disarmament Agency and then the State Department, I served as the Senior Advisor at the US–Soviet Nuclear and Space Talks in Geneva, Switzerland, during 1985–1988.

My most valuable contributions to national defense were made as a private individual, without remuneration. Certain of these initiatives incurred serious professional and some personal risk. Due in part to my efforts several misguided military endeavors were cancelled and more realistic, worthwhile programs were implemented. The B-70 Strategic Bomber Program and the Multiple Protective Shelter and Densepack basing schemes for the MX (later Peacekeeper) ICBM stand out as misguided developments whose demise was beneficial to the country.

Eliminating these developments saved taxpayers $10 billion or more in ill-considered defense expenditures, but this hardly suited the corporate participants.

At the same time I was a determined advocate of efforts that resulted in the development and production of more than 1,000 short-range attack missiles (SRAM) for the B-52, FB-111 and B–1 strategic bombers. I redefined the unique and necessary strategic bomber mission and the avionics bombing systems for the B-52 and B–1 were designed for the new mission. Twenty-one years of advocacy finally brought the B–1 Bomber to successful development, production and deployment in the mid-1980s.

I prepared a unique and comprehensive strategic force structure analysis in the mid-1970s that identified the nature and timing of the overwhelming Soviet strategic threat to the United States: "Air Force Study of US–USSR Strategic Forces (1975–1985)." This study, and briefings based on its findings, justified B–1 Bomber production and accelerated the MX ICBM development in 1976 under the Ford Administration. The subsequent 1980 Soviet threat analysis, "Evolution and Capabilities of US and USSR Strategic Forces (1975–2000)", facilitated the Reagan Administration's Strategic Forces Renewal Program. My work with the Administration during 1980–1983 and testimony to the 97th and the 98th Congress ensured agreements to produce 100 B–1 bombers, to base 100 MX ICBMs in the Minuteman ICBM silos, to develop and produce the Trident D-5 Submarine-Launched Ballistic Missile, and to develop a small land-mobile ICBM.

Then for four years I worked in Geneva, Switzerland, as the Senior Advisor to the Nuclear and Space Talks. During this period I prepared analyses of Soviet initiatives and a study that anticipated the systemic failure of the Soviet Union.

Most of my studies were classified as secret; but the passage of two or three decades allows sharing nearly forty years of work that may be of value to those interested in knowing more about the evolving balance of strategic military power. My more recently retired DIA and CIA friends have reviewed appropriate chapters of the text to ensure that I did not remember too much. Thus the following chronicle.

By law, all of my classified documents were left in Safe No. 1463759 in the S/DEL suite at the State Department, Room 2422 on June 2, 1989. The Air Force, DOD and other organizations mentioned are also likely to have archived a few of the 100 copies of the I/O-3 Final Report. These documents would be of possible interest to serious military scholars and eventually obtainable though the Freedom of Information Act.

The last chapter of this book contains an analysis on the current status of the struggle to maintain strategic nuclear parity and briefly reflects on the importance of individual service to our nation and society.

Chapter 1. Nike Air Defense

Before World War II even began to fade into the background, temporary allies reverted to their positions as rivals in the evolving contest for power in the world. Late in World War II, the US Army Frankford Arsenal had formulated the need and concept for a mobile, ground-based anti-aircraft weapon system that would provide much higher kill probabilities than conventional anti-aircraft artillery (AAA). Far greater effective range was needed to increase the intercept distance against nuclear-armed aircraft. And indeed, during the early years of the Cold War, the Russians were working to design and deploy long-range bomber aircraft.

As Soviet strategic bombers presented a growing threat to America and our allies, the Bell Telephone Laboratories (BTL) and the Western Electric Company were tasked with developing the Nike Anti-Aircraft Surface-to-Air Guided Missile System. Seventeen Bell System field engineers, including me, wrote the preliminary technical manuals on the Nike System and taught the first class of US Army officers at the White Sands Proving Ground (WSPG) in New Mexico. For three years, 1952–1954, the Nike Program included prototype and production model proof-testing, Nike Anti-Aircraft Battery site selection, and the initial deployments to defend our most important cities and defense installations. Within a decade the Soviet strategy began to focus more on ICBMs and SLBMs, presenting a much greater threat that we have failed to counter for the past forty years. Mutual deterrence has been our only defense.

The Bell Telephone Company developed and produced about half of the communications and electronic equipments used in World War II. Thus the company was selected as the sole source for the Nike System development and later production. The Douglas Aircraft Company (DAC) designed the missile, booster and launching equipment.

The Nike Acquisition Radar was nearly identical to the M-33 Search Radar developed earlier for AAA Battery target detection and designation. The monopulse Target Tracking Radar (TTR) derived the target's range and altitude from a single radar pulse and three-dimensional target velocity within a few seconds.

The radar operators could select targets from the acquisition radar and rapidly slew the target tracking radar to the general azimuth, elevation and range of the target aircraft. The TTR would immediately refine these measurements, lock onto, and automatically track the designated target.

The Missile Tracking Radar (MTR) was very similar to the TTR but was tuned to a slightly different frequency to avoid mutual interference. The MTR was locked onto the selected missile prior to launching. Throughout launching and flight to the target, the MTR transmitted continuously corrected commands that guided the missile to a pre-calculated intercept point and exploded the warhead close to the target.

This was the last generation of weapon systems before the digital computer. The Nike Central Computer was a large, complicated and very accurate analog computing machine. The critical components were complex and expensive precision potentiometers (variable resistors) that represented mathematical quantities or functions. The computer accepted all TTR and MTR data to calculate continuous missile guidance commands and the intercept solution.

The four vans in the Nike System included the Radar Control Trailer, Battery Control Trailer, Spares and Equipment Trailer, and the Launch Control Trailer. The latter van was central to the Missile Launching Area, separated by 3,000 feet or more from the radars in the Battery Control Area. The 25-foot vans were moved by heavy military trucks. Several motor-generator sets and frequency converters provided power for field operations.

Since the system was designed to be both air and road-transportable, aluminum and magnesium were used as the basic construction materials for the vans. These were the largest units that could be transported by our early 1950s airlift.

The Nike I Missile, later called the Ajax, was about 20 feet long and a foot in diameter. The missile was seated in an 18-foot booster of similar size. The booster contained a single-grain solid propellant that boosted the combination with accelerations up to 125 times the force of gravity; this boosted the missile to a speed nearing Mach 2.0 in a very few seconds.

The final version of Nike II or Nike Hercules was also under development. This larger missile used four of the Nike I boosters fastened together. It became operational after 1954. This increased the intercept range to over 70 miles at altitudes above any foreseeable aircraft and provided a single-round kill probability 100 times better than artillery.

After booster separation, the missile's liquid-fueled rocket motor provided powered flight. The missile fuel was JP-4, jet engine fuel, with red fuming nitric acid as the oxidizer. The two liquids were hypergolic, instantly igniting with explosive force when mixed.

The missile contained warheads connected by primacord that ensured essentially simultaneous explosions, providing a spherical pattern of fragments traveling at 7,000 feet per second. This speed allowed the fragments to go completely through an aircraft engine and caused momentary ignition of the aircraft's aluminum structure.

Following five years of development, the Nike I System destroyed a violently maneuvering drone target in November 1951 on the first controlled test flight. Soon after, with only a smoke marker for a warhead, the Nike missile destroyed a QB–17 drone in a head-on intercept by flying into the nose and out through the tail of the aircraft.

WSPG was 20 miles east of Las Cruces, across the Organ Mountains through the San Augustin Pass, at 5719 ft. The missile and artillery ranges included about 8,000 square miles of the 50-mile wide Tularosa Basin, including the White Sands National Monument. The Nike Guided Missile School was established there to instruct a cadre of about 20 Army officers, ranging from Warrant Officers through Majors, and several civilians from Fort Bliss and the Redstone Arsenal. These men later set up the Fort Bliss Guided Missile School to train the personnel needed to man the Nike installations throughout the United States.

V-2 Rockets, imported from Germany, were being launched at WSPG with the help of the scientists who originally invented them — they had preferred being captured by the Americans rather than the Soviets. The last 30-ton V-2 missile was to be fired at midnight during the late summer. We drove out to WSPG and hiked into the desert for a closer view of the launch. The V-2 was known to experience guidance failures that could cause it to rise a few hundred feet, tip over into low-altitude horizontal flight, and explode; Army MPs patrolled the area to keep spectators at some distance. Fortunately, this V-2 flew straight up for 50 miles.

Three field engineers went to Winston-Salem, NC in February 1953 to modify the Nike factory test specifications. We then installed the first production system at the US Army Redstone Arsenal at Huntsville, AL.

A month later, we sited the second Nike Production Model northeast of the Main Base at WSPG, in coordination with other Bell System, Douglas Aircraft, and WSPG personnel.

We found a unique way to align the target tracking radar: by locating a lone, distant cloud over the desert. There was always a buzzard or two riding the updraft beneath the cloud. Using the powerful boresighting telescope mounted on the radar, the antenna was pointed and the range gate adjusted to lock onto a single bird. With a monopulse radar that could automatically track a buzzard ten miles away, there was never any concern about our ability to track aircraft anywhere within our radar coverage.

During one missile test a few miles east of our installation, an Honest John short-range ballistic missile was launched, went straight up into the stratosphere, and then came more or less straight down. Our Nike vans shuddered from the impact a few hundred feet away. This was a practical demonstration of the coreolis effect: during the vertical flight up and down, the earth had rotated a few miles to the east, bringing our test site underneath the descending missile.

The Western Electric Company was to provide technical support for the growing Nike Field Engineering Force (FEF). Their headquarters was in Winston-Salem, NC, near the Nike manufacturing site at Burlington, NC.

In a practice which non-defense companies consider normal tactics of self-preservation, the FEF Management preferred individuals in management positions whose first consideration would be the company, not the customer. On vital defense programs this practice can be contrary to national interests, and in the following years many examples came to light.

Siting

The original design concept for the Nike System provided an air and ground-mobile anti-aircraft weapon for the defense of US Army forces in the field. Successful development and the growing threat of the Soviet Long Range Air Arm meant that the most urgent need for the Nike System would be permanent installations in and around major US cities and critical defense facilities such as the nuclear materials production plants at Hanford, WA.

The Bell System had provided little guidance to the Army on how to adapt Nike to permanent sites in urban areas. This inadequacy soon became apparent through many inquiries that came in from Army organizations and our own FEF. The Nike FEF Manager, Wally Ratcliff, lacked experience with siting Nike systems and general knowledge of the relevant government organizations; eventually, due to my background at Redstone Arsenal and WSPG, I was called upon to provide direct FEF technical assistance to the Army.

Given the growing magnitude of Nike siting problems and general confusion, Wally finally sent me to visit the Eastern Army Anti-Aircraft Command (EASTARAACOM) near Newburgh, NY. A Douglas Aircraft Company engineer, Walter Klass, and I met with Major General Hayden and his staff regarding Nike siting criteria. The General agreed to FEF participation in site preparation and put us in touch with WESTARAACOM, as well, in San Francisco. WESTARAACOM also requested our assistance. General Hayden asked Wally to assign me immediately as FEF engineer on his Headquarters Staff.

Initial visits to the Washington/Baltimore and New York AAA Brigade Headquarters indicated that many of the Nike siting problems originated at the top. Inadequate or conflicting guidance was coming down from the Army Office of the Chief of Engineers (OCE) in Washington and the Army Anti-Aircraft Headquarters in Colorado Springs. Existing plans also needed to reflect greater technical content, and flexibility in site selection and in the actual Nike Battery configuration(s). General Hayden gave us permission to work out these problems with OCE, AAA Headquarters, AAA Brigade Commanders and the US Army District Engineers in each defense area.

Since OCE and the subordinate commands had been working with these somewhat deficient plans for more than a year, many sites already had been selected and designed. Fortunately, few were actually under construction. The necessary 34 acres of land for each site had often been purchased through the exercise of the government's right of eminent domain. Consequently, many of the government principals were very resistant to changes in their plans, even if they had made serious mistakes in site selection and design.

The key to improving the Nike siting situation was the Army Office of the Chief of Engineers, and specifically Edward M. Rosenfeld, who was the principal architect for OCE. We revised the Nike siting plans and criteria and ensured each District Engineer was authorized to utilize advice afforded by the Bell System and Douglas field engineers assigned for that purpose.

The first prototype site with underground storage for the Nike missiles was being built near the prison at Lorton, VA. This was only a few miles from OCE, allowing close supervision of the construction. We visited the site several times during construction to ensure that there was compatibility between the equipment, cabling, missile handling equipment, adequate elevator capacity, and overall safety in storing the many tons of explosives in the missiles, warheads and boosters.

Ed designed bunkers with flanking revetments to ensure that any accidental blast would blow out the 3,000-square foot roof of the chamber and direct the explosion upward to minimize lateral surface damage. Three separate underground bunkers were to be provided in each Nike Launching Control Area. This made the missiles safer from sabotage, less obtrusive, and reduced the land area required for the launching area to 30 acres — an important consideration in suburban areas.

Having initiated the work on new site plans, we visited and consulted for authorities at all of the prospective Nike sites. These included about 130 sites already selected around the first 14 cities to be defended. Another engineer was trained to do similar work for WESTARAACOM. Walt Klass provided his expertise on the missile-related aspects of siting.

Having started these activities a year late, we made up for lost time by traveling over 50,000 miles in DC-3s and DC-4s and working 80 to 100 hours per week during the winter of 1953–1954.

FEF Headquarters was concerned to make sure we were putting company interests first. Wally attended one briefing to the military and civilian people in Cleveland, "to find out what I had been telling the Army." His only comment was that the briefing was too technical for the audience.

The fundamental document provided to the Army by BTL was an excellent operations research study by Donald Ling that described the best defense for a typical large city. The Nike Batteries were placed in a circle around the city to defend against an attack from any direction. The radius of the circle was determined by the lateral spacing necessary for the inter-battery fire support needed to deploy an effective defense outward from and over the entire city.

The second essential study for Nike siting was the Bonner–Brown analysis of each US city to be defended. These analyses defined the "Forties Contour Line." This line surrounded the portion of the city where an average-sized atomic bomb (the H-bomb was not yet developed) would kill 40,000 people and/or do $40,000,000 in structural damage.

New York

The latter study also defined the epicenter of the defense within the Forties Contour Line. For example, the central point of the New York City defense in 1952 was the east end of the Holland Tunnel. The large area to be defended required that the ring of Nike Batteries around New York be spaced outward 40,000 yards beyond the epicenter.

The Hudson River and New York Harbor required that the defenses begin with a double site at Sandy Hook (NJ) to cover the harbor. The sites circled clockwise to Fort Dix (NJ) northward, west of the Hudson River to Nyack (NY), eastward to the Rye (NY) reservoir, Livingston (on Long Island), and ending at Fort Tilden, near Coney Island (NY). The northernmost site, at Nyack, was placed on the highest hill on the west bank of the Hudson to defend against a low altitude attack coming down the river from the Bear Mountain Bridge.

Washington

Although New York was strategically the richest target, Washington as the capital city was the first priority for Nike defenses. The proximity of Baltimore produced overlapping circles of Nike sites. Consequently, a "banana-shaped" defense was to surround both cities under the command of Brigadier General T.V. Tayton.

Placing Nike sites among the rich and powerful could be difficult. Army staff cars were seen visiting prominent journalist Drew Pearson's farm west of Potomac, MD. When the inspection party came back to the Brigade Headquarters, three congressmen had already called, telling the Army to select a different location for that site.

To ensure inter-battery fire support, the 25-mile range of the Nike I (Ajax) Missile required an initial siting of batteries along a circle that, a dozen years later, became the Capital Beltway, Interstate 495. Later, an outer ring of missile sites was planned for deployment as the Nike Hercules Missile became available. The 70-mile range of the larger missile allowed inter-battery fire support in all directions. This was always an important consideration in design of the overall defenses because a Nike battery could engage only one target at a time.

Some of the Nike sites with primary fields of fire over water or areas of low population would eventually have Nike Hercules missiles armed with nuclear warheads. One such site was west of Potomac, MD, in the outer ring. Without giving a full explanation, the Corps of Engineers attempted to buy the entire farm. The gentleman farmer refused and as a consequence had an underground nuclear arsenal a half mile from his house for the next ten years.

There were trade-offs and complexities to be weighed when selecting sites. Don Ling recommended that the launchers be placed forward of the Battery Control Area in the 120-degree primary field of fire to allow the earliest possible engagement of approaching aircraft. The idea of placing the launchers behind the Battery Control Area for better inter-battery fire support was a specious plan that required more Nike batteries to provide the same degree of defense. The greater range of the Nike Hercules removed any logical basis for this tactic. At the same time, military and civilian authorities siting Nike around Philadelphia were emphasizing "Over-the-Shoulder" sites with launchers behind the Battery Control Area. We later inspected an over-the-shoulder site and found they had placed the four-acre Battery Control Area in the middle of the battery's booster disposal zone.

San Francisco

This city presented the most difficult siting problems in the United States. Fortunately, Fort Baker was the headquarters for WESTARAACOM, giving the city one of the best Guided Missile Officers (GMOs) that we encountered in this arduous tour. The Colonel had done a remarkably good job of placing the sites in ideal but very challenging locations. Our task was primarily to help him adapt the Nike System to the topography, particularly the double site on Angel Island in the Bay and another at Fort Kronkite atop a mountain at the northern end of the Golden Gate.

The mountain peak at Fort Kronkite, overlooking the Pacific Ocean and the Golden Gate Bridge, had the best view of any Nike site in the United States. This site also had superb radar coverage in every direction — if we could adapt the Battery Control Area to this precipitous location.

To allow line-of-sight and a distance of 250 feet between the tracking radars it was necessary to build a concrete platform to elevate the Missile Tracking Radar. Since the boresighting mast would be 60 feet below the radars instead of 60 feet above them, we made a field change to permit the corner reflector at the top of the mast to tilt upward as well as down. This change was also necessary for a problem site on Hog Island in Boston Harbor.

The three city defenses described above were typical of the problems we resolved for the Army. The other cities included Boston, Norfolk (to protect the Atlantic Fleet), Buffalo/Niagara Falls (Chemical factories), Chicago, Detroit, Cleveland, Pittsburgh and Los Angeles.

Our recommendations in siting and site design improved the effectiveness and reduced the cost of the overall Nike deployment. The on-site consulting saved the Army several million dollars in current and later construction costs by relocating 12 per cent of the sites and reorienting 20 percent of all sites. This included the avoidance of unnecessary forest clearance, earth moving, and construction of many radar tower installations. Local topography required a few over-the-shoulder site selections, but most of the rest were eliminated.

Overall, the Army estimated that they could now reduce their Nike procurement for continental defense by 15 battalions or 60 Nike batteries. This would save about $250,000,000 in equipment costs and a similar amount in 10-year maintenance and operational expenses. Half a billion dollars was a lot more in 1954 than it is today: The annual defense budget was then about $50 billion, having recently increased three-fold due to the Korean War.

EASTARAACOM offered to extend our contract for three months on similar assignments — recognition that the consulting input had made a significant contribution to the defense of the country.

During the five months of solving siting problems, BTL-Whippany provided technical expertise. Phil Thayer, the BTL engineering manager for the Nike Development Program, specifically warned that consulting for the Army was limited to technical advice and should not include the tactical aspects of the Nike Deployment. He added, "The responsibility of the Bell System is to provide the Army with an effective weapon system, not to advise them on how to deploy the weapon — that is their business." In other words, the Bell System's position was that if the US Army erred in deploying the Nike System, in a way that led them to contract for more than they needed — that was their choice. Bell expected employees to avoid discussing the subject. I therefore refrained from touching on tactical deployment unless I was alone with a given GMO. These able officers resolved any tactical deployment problems by themselves.

The lack of support for the Army's siting efforts and Wally Ratcliff's unusual visit to our Cleveland meeting likely were related to Mr. Thayer's concerns. This suspicion was confirmed during my first meeting with Wally after the EASTARAACOM assignment was completed. I asked FEF Headquarters in late March 1954 whether I should continue working for EASTARAACOM for another three months, as the Army had requested. Wally vigorously denied the request, and forcefully suggested other positions for me in the Bell System.

Bell Telephone Laboratories

Fortunately, Phil Thayer at BTL was aware of the Army's needs and invited me to BTL-Whippany to continue working on Nike — and by the way ensuring that liaisons with Nike were more closely supervised. Management was nervous about my role, but our effective efforts on behalf of the Army were well known and my continued presence had to be considered in light of customer relations.

I began working as a de facto member of the BTL Technical Staff. BTL was then the foremost engineering laboratory in the United States, providing an excellent learning experience. The first assignment was to write up everything I knew and had experienced regarding Nike siting, including a summary of work in each of the 14 cities. This information was then shared by Colonel Crane, who was still responsible for the Army liaison office at BTL, with the various Army commands and later through channels to the German, Netherlands and Norwegian governments as a guide for further deployments both in the US and NATO.

Special Nike problems included the design of sun shields needed to surround tower-mounted tracking radars to preventing differential heating that would misalign the radars. These structures later surrounded the tower-mounted radars at Key West when Nike was deployed there during and after the Cuban Missile Crisis.

Another task was the design of sound baffles for the underground bunkers beneath the missile launchers. During Nike launches at White Sands, lift-off entailed a gut-rattling, deafening 130-decibel noise level experienced a half mile from the launcher.

I made several trips to OCE, providing information to Ed Rosenfeld. After our last visit, he was transferred to the Redstone Arsenal. During his thirty years there, Ed selected the Kwajelin Atoll site for the Army's down-range anti-ballistic missile test facility. He was responsible for the design of the entire installation. Forty-three years later, Ed admitted that he had paid a high price for revising the Nike site plans. The transfer to Huntsville had not been voluntary and it ended a very promising architectural career at OCE Headquarters.

In late June, I asked Phil Thayer to transfer me to BTL as a permanent member of the technical staff, only to learn that I had "become a controversial personality." No one ever mentioned directly the $250,000,000 of lost revenue to Bell Systems when the Army improved its siting plans for the Nike System.

Chapter 2. IBM Military Programs

During 1942–1945, the International Business Machines Corporation converted much of their manufacturing capability to war production. After World War II, they rapidly returned to the much more profitable business of office machine design, development, manufacture and services.

When Thomas J. Watson, Jr. (a B-24 pilot in World War II) ascended to the presidency, with the perception of a growing threat from the Soviet Union, and the continuing Korean War, IBM again participated in the defense industry. They became an alternate supplier of bombing and navigation systems (BNS) for strategic bomber aircraft.

In 1953–1954, IBM established a development capability for analog bombing and navigation systems intended primarily for the new B-52. The Airborne Computer Laboratory in Vestal, NY, was the new engineering facility for BNS and a few other military programs.

IBM also accepted sole-source development and manufacturing responsibility for the Semi-Automatic Ground Environment System (SAGE). The system consisted of an immense ground-based computer and communications complex to coordinate air defense guided missile and aircraft interceptors over large areas. Nationwide and Canadian deployments were planned. SAGE development was based in Kingston, NY, near Poughkeepsie, for substantial support from the IBM commercial laboratories.

IBM preferred to hire the best new engineering graduates and have them evolve in the established IBM culture. However, the new and diverse skills needed to rapidly establish military systems development required some exceptions to this practice. During 1954, Coke machines were accepted and the IBM song books went out the back door. Corporate paternalism was still firmly established, very beneficial but less smothering.

Hardware Development

ACL Management had responsibility for design of the main SAGE display console in support of the IBM-Kingston facility, but they were months behind schedule.

My first assignment was to redesign the deflection circuits for the very large charactron display tube.

The charactron, developed by General Dynamics, was one of the largest extant electromagnetic cathode-ray tubes. The tube was about 30 inches in diameter and intended to display large areas of radar coverage. It contained a metal grid with holes for each alpha-numeric character that was displayed next to radar targets. A computer-controlled electron beam from the cathode was rapidly sequenced through this grid to display the selected numbers and letters on the display screen.

Building F at the Massachusetts Institute of Technology housed the first developmental SAGE computer and displays. When the manufacturing prototype of the charactron display was installed there, in mid-February 1955, the unit did not function properly. I spent several days at MIT and corrected obvious difficulties and made circuit changes that had tested successfully in the Vestal laboratory.

The core of IBM's new military hardware development was the MA-2 Bombing-Navigation System for the B-52. The BNS equipment was being developed and manufactured at several locations in Binghamton, Johnson City and Endicott, NY.

In 1955, the hundreds of analog computers and other subassemblies in the BNS were packaged in metal "beer cans," mounted in cabinets and interconnected through laced wiring networks (cable harnesses). IBM had many start-up problems with their new BNS business and missed deliveries to the first A-through-D Models of the B-52s.

The Air Force badly needed another manufacturer to augment or replace their other BNS producers, the Bell System and Sperry-Rand Corporation. The presumably more desirable B-70 Program was also on their business horizon.

The ACL in Vestal was established to support the IBM MA-2 development and manufacture, but it was intended primarily to develop new technology for strategic bomber avionics. IBM had no peers in the newly emerging business of developing and manufacturing large digital computers. Consequently, their first effort at ACL would be to develop an airborne digital computer to replace and greatly improve upon the existing analog, beer-can computer technology.

ACL Management conceived the DINABOC (Digital Navigation and Bombing Computer) project to centralize all of the primary computations required for navigation and weapon delivery. The central computer was to receive and analyze the optical, video, electrical, range and angle data from the dozen or more sensor subsystems and continually compute the solution to the bombing equations for weapon release.

Little work had been done on the problems of converting the analog sensor data to the digital form required by the central computer and for feedback to certain of the sensors. Throughout the companies supporting the military establishment, from GE to BTL and IBM, behind-schedule, time-urgent, complex projects and difficult conditions were to be expected.

The specification and design of all the analog-to-digital converters for the experimental DINABOC system took me through 1956. This equipment later evolved into a digital computer controlling an entire B-52 BNS. The laboratory included a landmass simulator to synthesize the radar returns of bombing missions. This installation was later modified for the B-70 BNS configuration. The laboratory served as a useful engineering and sales tool for the B-52 and B-70 avionics for the next several years.

B-70 Bombing-Navigation System

The new B-58 supersonic bomber did not have the range and payload to warrant replacement of the B-52s. The next US Air Force strategic bomber was to have sustained supersonic flight over intercontinental distances with multiple weapons. North American Aviation (NAA) came up with a remarkably advanced design for Weapon System 110A, later called the B-70 Valkyrie.

Stainless steel and titanium construction, six very powerful GE engines, a unique fuselage design, partially folding delta wings, and canard control surfaces would allow the B-70 to fly several thousand miles at altitudes over 60,000 feet while maintaining a continuous speed of Mach 3.0 — over 30 percent faster than could be sustained by aluminum aircraft. The Air Force was run by pilots who believed new technology, greater speed, and higher altitudes necessarily produced a more effective strategic bomber — and it would be really be fun to fly.

The avionics contract was awarded to IBM in February 1956. The B-70 BNMGS was far more advanced in technology, sophistication and performance than the B-52 or B-58 avionics. Some common subsystems were used, allowing the adaptation of 15 B-52 data converters to the B-70. Nine more analog-to-digital electromechanical converters of unique design were needed, for a total of 24 in the bombing-navigation portion of the avionics. I designed and oversaw the fabrication of these units until February 1957.

The next assignment was to design and provide technical supervision for of the input-output converters for the new Doppler radar and the stellar-monitored inertial platform. The latter unit and the associated complex subsystems had been developed by the NAA Autonetics Division for the intercontinental, nuclear-armed Snark Cruise Missle.

We also designed a B-70 variant for the DINABOC laboratory facility to provide new converters for the latest AN/APN-96 Doppler Radar. This was the first reversible analog-to-digital electromechanical data converter, a dual conversion device for the Doppler-derived drift angle used to improve the aircraft's true heading data._

The gyros and accelerometers of the N2C-I Inertial Platform were the most accurate available in the mid-1950s — comparable to those used in first generation ICBMs. The platform provided three-dimensional velocity data, inertial heading, and their rates of change. These data allowed precise inertial navigation.

The star-tracking telescope had to be mounted onto the gimbaled inertial platform assembly to ensure that both devices were working in exactly the same mechanical coordinates. The tracker would simultaneously track two selected stars among the 70 brightest visible from the earth's northern hemisphere. Analog, preprogrammed, punched paper tapes were used in the Snark Cruise Missile application to select stars and provide initial pointing angles for star acquisition after the rough alignment of the N2C-I Platform. We eliminated these tapes by storing the star tables in the central computer.

The star-tracker necessarily worked with sidereal (star) time rather than our solar time. The resultant very precise position data was referenced to inertial space rather than earth coordinates._

The platform azimuth data required a two-speed analog-to-digital data converter. The stellar telescope pointing accuracies, measured in arc-seconds over a range of two radians, necessitated a three-speed synchro/converter system. The star-angle converters were also unique and patentable, but due to the narrow applications of both the Doppler radar and star-tracker converters, IBM published the designs rather than patent them.

IBM's rapid growth in the strategic avionics business required building a major manufacturing and engineering installation in Owego, NY, a small town on the Susquehanna River twenty miles west of Binghamton.

A 600-acre hillside site a mile east of Owego provided 500,000 square feet of floor space and would eventually accommodate 5,000 employees, twice the number of people who lived in the town. Construction was completed in December 1957.

The usual dichotomy emerged wherein advanced technical work was managed by people who lacked the ability to do the work themselves and "gee-whiz" solutions were chased in hope of superseding plain hard work. At one point, a professor from UCLA was hired to help design a new analog-to-digital converter to replace my converter designs. They spent six months and a lot of money to develop this alternative; in a few hours my technical evaluation showed the design to be fundamentally flawed.

Next I was given technical responsibility for the complete integration of the N2C-I Stellar-Inertial subsystems for the B-70 Flight Test BNMGS, tying together the specification, design, fabrication and test of all the remaining data conversion hardware. Three documents defining the overall computation requirements for the electromechanical and electronic conversion equipment were produced in September, 1957.

The Owego facility architects had not been advised of the need for a star-tracker laboratory. My last task on the B-70 was to design a 1600-square-foot one-story adjunct to Engineering Building I. This included a roll-back roof and cable trenches in the concrete floor that would later be used to interconnect the completed B-70 BNMGS for pre-flight testing.

A "stable table" column for mounting the inertial platform was sunk to the bedrock and isolated from the laboratory floor. The structure prevented platform vibrations and tracking errors when freight trains passed by about 4,000 feet from the facility. Systems Research also made their contribution, noting that, historically, Owego usually had only one cloud-free day in February.

B-70 Development Engineering was formed to further design the central digital computer and improved versions of the 35 engineering prototype converters. The adequacy of the original designs, difficulty in staffing skilled people, and the early compressed schedules for B-70 development precluded the development model redesigns. Later on, as B-70 fortunes faded, schedules were more relaxed but money was no longer available. Everything that I had designed in the previous two years later flew as the final B-70 BNMGS Flight Test System.

IBM famously provided a "Think" sign for everyone's desk to encourage individual initiative. But IBM management in Owego and Kingston had not yet thought about the long-term military trends that could affect their programs.

Immersed as I was in working to help IBM win and perform on the B-70 BNMGS Program, I still had a great private concern about B-70 operational survival. Was the Air Force properly assessing the potential lethality of the Soviet Air Defense System at high altitudes? The Nike Guided Missile System at WSPG ensured that the days were numbered for bomber penetration at altitudes from 1,000 to 70,000 feet. Further, with Nike Hercules armed with small nuclear weapons, the entire upper atmosphere could be rendered impenetrable for manned or unmanned aircraft. Could the Soviet Union be far behind? I hoped to find the information that would allow a proper analysis of future threats.

In the new Owego facility, Art Cooper was now manager of a new function called Advanced Systems Research. I encouraged him to consider the need for informed, systematic analysis of major military trends affecting heavy bombers and the SAGE Air Defense System in Kingston. Cooper had me transferred to his department to do applied research planning and pre-program weapon system analysis.

During four years as a major defense contractor, IBM had not yet explored the data resources and future plans of the Air Research and Development Command (ARDC) at Andrews Air Force Base near Washington, DC or the subordinate Aeronautical Systems Division (ASD) at Wright-Patterson Air Force Base in Dayton, OH. Since the funding and contracts for new aircraft and avionics developments came from ASD, I established a working relationship for IBM with the very capable ASD planning staff.

They provided some secret planning documents that could be released to industry. In return, we promised to provide them with the results of our thinking and an IBM mathematician to help with some of their statistical analysis methodology. Dr. Wouter Vanderkulk, employed at the Owego facility, was our best applied mathematician. He solved several of their problems within a few days.

As IBM representative, ASD Plans referred me to the ARDC Headquarters Plans Office which maintained the Systems Requirements Board that conceptually defined the advanced weapon systems under consideration by the Air Force. IBM had never reviewed the SR Board.

ASD Plans also introduced me to the Air Force Technical Intelligence Center at nearby Patterson Air Force Base. ATIC was the Air Force intelligence organization that collected, correlated, analyzed and reported all of the technical intelligence data provided by the Air Force and many other intelligence agencies.

Practically the entire ATIC effort was directed toward the Soviet Union to report on current and projected USSR offensive and air defense weapon systems. In 1957, only one officer, in a staff of 100 or more military and civilian scientists, was needed to monitor the technical and military programs of the Peoples Republic of China. Before Sino–Soviet relations were badly strained in 1962, practically all PRC technical improvements came from the USSR — several years after the Russian deployments.

The "mother lode" of information had at last been found. ATIC had the data that was needed to perform an independent analysis of the major trends in US and Soviet weapon system developments.

The Aircraft Branch at ATIC was run by Lt. Col. J.J. (Jack) Henderson, who became a life-long friend. Jack Henderson was P-38 pilot in Europe during World War II, flying fighter, ground attack, and reconnaissance missions. While in Europe, for whatever reasons, he enjoyed the confidence of General Eisenhower. This relationship continued when Eisenhower became president. He later had a similar relationship with President Lyndon Johnson. He was also the personal pilot for the general(s) commanding WPAFB, whenever they used their DC-6 for world tours.

I spent the next several months studying the many ATIC intelligence documents describing the massive Soviet Air Defense System's surface-to-air missile systems (SAMS) and interceptor aircraft. Their technology and developments would eventually preclude any manned bomber penetration at altitudes above 1,000 feet. This confirmed the vulnerability of the B-70 Weapon System — years before the intended 1964 deployment.

The Soviet Rocket Forces were similarly increasing their intercontinental ballistic missiles in both quality and quantity; at this rate, by the mid-1960s they would have enough sufficiently-accurate missiles to destroy any unhardened military or industrial installation, 50 US Strategic Air Command (SAC) bases, and a hundred of the largest US cities.

The low accuracy of Soviet ICBMs in the late 1950s (3,000 to 6,000 feet), was more than compensated for by their huge warheads, having a yield of 20 megatons or more on some models. This growing capability would allow the Soviets to eliminate our SAGE air defense installations several hours before their bomber fleet arrived in the United States.

By late 1957, my analyses showed that IBM's two main military programs, the B-70 and SAGE, were based on fundamentally fallacious assumptions and would be subject to cancellation whenever the real trends gained wider recognition. IBM at that time maintained a full-employment policy and had several thousand people committed to these programs.

Improving US and NATO air defenses and providing cost benefits to the government through improved Nike siting had cost me my position at Bell. The new IBM circumstances were crucial but less urgent. I still felt strongly that the interests of our nation should come before those of my employer.

The IBM situation differed from that at Bell Systems in that the B-70 and SAGE were still in development and the senior managers were less experienced and guided by higher principles. Before the expensive production phases began, I thought, the two programs could be reduced or eliminated by propagating valid concerns about their inherent operational vulnerabilities. At the end of 1957, the situation was not generally or clearly understood; but in subsequent years I encountered formidable opposition to my individual initiatives along these lines.

Using First Principles analysis, the following conclusions could be drawn: (1) Ballistic missiles would eventually become a much more effective and less costly means of strategic intercontinental weapon delivery against targets of known location; (2) Strategic bombers would be unable to penetrate the Soviet Union at high altitudes; (3) The B-70 could not fly a militarily significant distance at low altitudes, and (4) The B-52, although designed for high altitude operations, could also fly at altitudes below 1,000 feet for several hundred miles after nearing the borders of the USSR.

To justify the continuing existence of the manned strategic bomber, a unique and necessary mission for it had to be defined. Its unique characteristic was that the aircrew could detect and attack targets of unknown or vaguely known location in real time during a single mission. In later years, when strategic missile forces would be numerous, they could also do damage assessment of a previous missile (or aircraft) strike on a target and, if necessary, immediately re-strike that target. As the number and reliability of ICBMs and sea-launched ballistic missiles (SLBMs) increased and their accuracy improved to a miss-distance of 1,000 feet or less, early damage assessment would only be needed against hardened targets such as missile silos.

My analyses determined that low altitude penetration, bomb damage assessment (BDA), and prompt re-strike against hardened strategic targets would become the unique and necessary mission of any manned bomber. But none of our then-current bomber force of B-36s, B-47s, B-52s, B-58s, or even the intended B-70 were equipped for or capable of performing this mission.

The question of uniqueness and necessity governed my analytical approach to weapon system design and evaluation for the next 25 years.

Demise of the B-70

During an early review of the ARDC System Requirements Board in January 1958, SR-163 indicated Air Force interest in armed reconnaissance aircraft. Art Cooper approved an internal a five-month study of the B-70 as a reconnaissance–strike aircraft, or RS-70 (without recourse to North American Aviation, our prime contractor). This allowed IBM to formally present our views to the Air Force for a strategic armed reconnaissance mission for the B-70 in May 1958.

My study of the future of manned strategic aircraft, "Environmental Basis for Future Weapon System Development," was a logical, definitive argument that predicted the armed reconnaissance mission and the potential cutback or cancellation of the B-70. This prediction was later upheld but it hardly suited the Owego senior management, since the Owego facility had been built to develop and produce the B-70 BNMGS. After another year of study, I presented an improved paper in May 1959. Senior management quietly formulated "Operation Overnight," a plan to reassign 1,450 B-70 employees to other jobs if the program was cancelled as predicted.

My prior agreement required that IBM provide any potentially useful work product to ATIC in return for their intelligence documents. Later in 1959, the B-70 briefing and report were given to Lt. Colonel Henderson and his superior, Colonel Cruishank. Jack was particularly interested in the briefing charts and asked to retain them.

In late November 1959, NAA-Los Angeles was devastated by the "Thanksgiving Massacre." President Eisenhower announced the abrupt and complete cancellation of the B-70. He later agreed to build only two prototype aircraft to "prove the technology." The IBM B-70 BNMGS Program was eliminated.

Jack Henderson acknowledged a month later that he and Colonel Cruishank had briefed President Eisenhower with my story, enhanced by their top secret data on the subject. The President's confidence in Jack, earned in World War II, permitted this out-of-channels communication. Ike's warning, late in his presidency, about the "Military–Industrial Complex" may have been influenced by the B-70 episode.

The cancellation of the B-70 Program was a significant contribution to national defense but a major blow to defense contractors. The B-70 would have been terminated or cut back anyway, in later years, as the facts became more obvious. However, that would probably not have happened before several billion dollars more had been expended.

The Air Force was now forced to reconsider the mission of maintaining the atmospheric weapon system threat to deter the Soviet Union. I spent the next 20 years helping them improve the B-52 to solve the interim problem posed by vacating the B-70 approach and the longer term problem of developing and deploying an entirely new strategic bomber.

In 1958, Art Cooper became Manager of the Advanced Systems Research Department. Monroe Dickinson took over dynamic analysis for the B-70. My proven converter designs always ensured static stability, but computer and subsystem dynamic interaction had to be assured by software design, analysis and system tests.

After six months it became obvious that my independent systems analysis work for Art Cooper was more useful and broader than the comparable work in Dickinson's department. Dickinson insisted that the work belonged in his department, and in September 1959 I was transferred.

Cooper and I maintained a special relationship and in December 1959 we went to a government/industry meeting at the North American Defense Command (NORAD) in Colorado Springs. The trip was an opportunity to have a private, classified, late-evening discussion with some senior people from the Department of Defense. With the B-70 BNMGS Program canceled and nothing to lose, we gave them the unvarnished internal IBM briefing on the anticipated B-70 operational environment — the same version furnished to Jack Henderson several months earlier. If they proceeded as planned without IBM's B-70 BNMGS development, the Air Force would have nothing but a high performance airframe with negligible military potential. We asked for $15,000,000 to finish the Flight Test BNMGS and one prototype system.

A pro forma meeting of DOD people at IBM-Owego on December 21, 1959 (arranged through our contacts at Colorado Springs) had much the same content, but a less frank discussion. Our BNMGS program was soon partially reinstated with $14,500,000. This allowed IBM to keep most of the B-70 engineering group intact. However, IBM was no longer taking $150,000,000 to develop the wrong bombing system for the wrong aircraft.

Due to confusion, militant self-interests, and the extensive funding at stake, a great deal of irresponsible information was generated in the next three years, aimed at re-establishing the B-70 Program. While IBM contributed to understanding of the problem, this was generally done in the foregoing technical and operational framework, thus avoiding any irresponsible IBM positions on a matter of national interest. Having correctly predicted the failure of the B-70 Program two years earlier and helping to salvage a skeleton IBM BNMGS Program, my actions were professionally unassailable — but were very unpopular with management.

My efforts to explain the fundamental problems of the B-70 had been carefully kept under wraps in IBM. After more than two years, the IBM B-70 BNMGS program manager reluctantly arranged a briefing to the B-70 Weapon System Project Office (WSPO) at ASD-Dayton on August 11, 1960. To his amazement, the story was well received and later was directly utilized by the WSPO Chief. The WSPO General asked us to consult with the Pentagon Air Staff Requirements Office responsible for Air Force advocacy of the B-70. A colonel on the Air Staff asked me to help him prepare for his briefing to General Curtis LeMay.

I wrote "B-70 Environment and Utility" in two days. The colonel was impressed by the effort but appalled by the conclusions. After this episode, he wisely moved on to other Air Staff responsibilities related to B-52 operations. Colonel David C. Jones later became a four-star general, Air Force chief of staff, and chairman of the joint chiefs in the early 1980s.

The B-70 paper was approved by Art Cooper and was furnished only to the B-70 WSPO. It was later approved by IBM Division and Corporate management, but it ran contrary to NAA views and self-interest regarding the B-70. Accordingly, IBM never responded to repeated requests from NAA for the briefing. IBM was big on customer sensitivity and NAA had an immense installation of IBM commercial computers that generated far more revenue than our reduced BNMGS program. However, on January 16, 1961, NAA Vice Presidents Schilling and Rice made their first visit IBM-Owego. They refused the elaborate review of the facility and the B-70 BNMGS that had been planned and demanded to hear my B-70 briefing. They were a hostile audience and said that they had heard it all before and did not accept the arguments. They promptly returned to Los Angeles.

After that, Owego senior management was noticeably weary of their excessively accurate seer on the future of the bombing business. And they weren't even aware of my prior role in the B-70 cancellation through ATIC.

Art Cooper sent the B-70 paper to IBM Corporate Vice President and Chief Scientist Dr. M. Piore, who apparently returned the copy to Art with a note saying, "Thank you for Carpenter's paper, I liked it fine." That likely extended my tenure at IBM.

My 1958 analysis of the Soviet strategic offensive forces indicated that their first-generation ICBMs could readily destroy permanent above-ground air defense installations. Consequently, one could assume that the huge, unhardened SAGE air defense control centers being designed at IBM-Kingston would be destroyed several hours before any arrival of the Soviet Long Range Air Arm. Briefings to this effect at Kingston were initially ignored.

IBM-Kingston had contracted with the Cornell Aeronautical Laboratories (CAL) to estimate and evaluate the Soviet bomber threat to the United States. Bob Crago, general manager and chief engineer for the facility, provided me with the CAL documents for review. While ignoring the Soviet ICBM threat to SAGE installations, CAL touted the SAGE System by overstating the bomber threat by 200 to 300 percent.

Art Cooper was advised that the CAL force structure exaggerations went far beyond the official Air Force bomber threat developed by the Air Technical Intelligence Center (ATIC). I drafted a letter to Crago but Cooper insisted that it be reviewed with ATIC. Their policy was never to critique such reports, but they were clearly amused. The letter was consistent with their estimates, so Cooper sent it to IBM-Kingston. We were spared the subsequent Kingston–CAL dialogue.

An annual briefing was also given to IBM-Kingston on the evolving strategic offensive and defensive facts and projections. Following the CAL incident, they were more receptive. When the initial 1960 Air Force cutback in SAGE occurred, the FSD Manager of Marketing asked for my assistance in defining other military business endeavors for the Kingston facility.

I visited Convair-San Diego regarding possible teaming with IBM-Kingston. Kingston could do the ground-based data processing for their WIZARD Anti-ICBM Program. IBM-Owego, now renamed the Space Guidance Center (SGC), could provide the on-board missile guidance. Convair gave us their missile guidance equations and SGC rapidly outlined the required computer design. IBM-Kingston handled the teaming arrangement and ground-based computation requirements.

The Bell System, using the BTL group that had designed the NIKE Air Defense System, was the leading competitor with their Nike Zeus Anti-ICBM System. Soon after the Convair-IBM Team got underway, DOD cancelled the WIZARD AICBM Program to proceed singularly with Nike Zeus. The Bell System built one such AICBM installation in North Dakota that the Army operated for a few years.

The ORION Program was intended to orbit huge, militarily significant payloads by sequentially exploding small nuclear charges against a "pusher plate" on the bottom of a massive multi-storied structure. This orbiting space station would be equipped to perform strategic surveillance and both offensive and defensive military operations.

The Air Force Special Weapons Center (AFSWC) at Kirtland AFB, NM, conducted this formative program as a series of separate but interrelated research efforts. During 1959–1962 we maintained a continuous IBM liaison with the project office because the Orion concept seemed to be the only foreseeable way to orbit enough mass with one vehicle to provide a militarily significant manned space weapon system. The ORION program activities were closely held and politically unacceptable.

AFSWC had a very secure room containing many types of nuclear bombs. Examining a 19-megaton hydrogen bomb (smaller than a Volkswagon) was analogous to a caveman looking at the first wheel. Both the Soviets and the US considered that bigger was better in the early 1960s. The Soviets eventually exploded a weapon in the upper atmosphere above Novaya Zemlya that yielded 63 megatons. If it had offered any military utility, we could have topped that.

MIDAS was to be a global system of orbiting satellites capable of detecting and tracking the infrared plumes of ICBMs and SLBMs when they reached altitudes above the earth's atmosphere. Such a system was essential to provide tactical warning (tens of minutes) at the beginning of a ballistic missile attack.

Analysis of the environment for future strategic weapon systems led to the conclusion that MIDAS was one of the few long-term, militarily-valid programs in the IBM-Kingston business area. Our initial discussions in DOD, USAF, ARDC and the Air Force Ballistic Missile Systems Division stressed IBM's capability to perform the large ground-based data processing tasks necessary for an effective MIDAS System. IBM-Kingston people and others were relocated to Bethesda and Gaithersburg, MD. These endeavors were developed into substantial programs that lasted for more than 20 years.

The SAGE Program faded away as anticipated. At least one SAGE site was located near the airport at Syracuse, NY. A later hardened site was planned for Canada, but it may not have been constructed.

Civil Defense and Shelters

International fear and suspicion continued to grow during the 1950s. The Soviet Union, still burdened by post-war exhaustion, had recovered sufficiently to impose the Berlin Blockade from mid-1948 to mid-1949, leading the US to respond with the Berlin Airlift. The Soviets developed the atom bomb by 1949 and copied the B-29 Superfortress in 1953, becoming an intercontinental nuclear threat to the United States.

The USSR achieved the hydrogen bomb in 1954 and exerted political and military influence beyond their borders for the first time during the Suez Crisis in 1956. A clear and consistent pattern of Soviet aggression and expansion had been established.

As a logical and necessary adjunct to strategic Soviet military capabilities in offensive and air defense weapons, they established a massive civil defense program to protect their leadership and civilian population. For example, the elaborate Moscow subway system is 200 feet further below ground than required simply for urban transportation. This massive network was constructed using special measures, such as blast-resistant doors to seal the tunnels, to protect their population from nuclear weapons.

Beginning under Stalin, the Soviets also built many miles of underground structures more than a half-mile beneath Moscow and other areas. These were structures intended to protect the Soviet political and military leadership in the event of nuclear war.

The US civil defense program could be evaluated as between farcical and futile in comparison to that of the USSR. Had the tragedy of general nuclear war occurred, their preparations would have given them decided advantages in the survival and recovery of their nation.

Analyzing the 1958 ATIC quantitative projections of the Soviet strategic forces and nuclear weapons gave me great personal concerns. Various terrifying scenarios came to seem more and more credible. A five-megaton bomb being dropped on Binghamton from a Soviet Badger (similar to our B-47 medium bomber) appeared to be the most probable threat during the early 1960s.

A five-megaton nuclear weapon could dig a crater 3700 feet wide and produce a fireball three miles in diameter. The blast would produce overpressures from a low altitude burst that would shatter our house and/or burn what was left standing — at least seven miles from the probable Ground Zero. Local and longer-term radioactive fallout would be difficult to predict, but it would be rendered harmless beneath two feet of concrete.

We decided that the risk was real and enduring. A simple basement fallout shelter would not ensure our survival; this would require serious work. To derive some functional benefit from the shelter, we decided to build an attached underground shelter with a family room above. We could enter the shelter through a new doorway in the existing basement wall. This plan eventually produced a 17' x 19' family room on a floor two feet thick and weighing 30 tons. Planning for the shelter was based on Federal Civil Defense Administration Technical Bulletin TB-5-3, published in May 1958. A 12' x 12' x 6.5' underground shelter was designed with a ten-inch structural cap, a fiberglass thermal barrier, and 14-inch concrete filler cap.

The local civil defense officials (both of them) were supportive as were potential building contractors. The Binghamton Steel Company and John Dakin, the Broome County Senior Engineer, helped to complete the structural design. These organizations were very interested because this was the first serious shelter construction effort in the Binghamton area. The total structure was to weigh 60 tons. Construction was completed in the fall of 1959.

Threats we sought to protect ourselves against included the flash, prompt radiation, blast, heat, firestorm, local anarchy, longer-term fallout, possible bacterial and chemical agents, imported anarchy, thirst, hunger and disease. We estimated these dangers would variously exist for 30 days before recovery could begin. We did not give much thought to the recovery, because a failure to solve all of the initial problems would preclude any recovery.

We intended to cope with the initial flash, prompt radiation, blast and heat by being at home. The time of day and the time of year for a premeditated or inadvertent attack eventually became more predictable. For then-classified reasons, a late evening in October was thought to be the most dangerous time. Since Binghamton was a secondary objective, perhaps 70th on the Soviet list of urban/industrial targets, we would have tactical warning. (The family and neighbors were advised not to look skyward if an attack threatened, to avoid being blinded by an early atmospheric nuclear explosion.)

If it happened, we assumed we would lose our house but the thermal barrier in the shelter roof would protect against the firestorm. Two large oxygen tanks would allow us to live without an external air supply for twelve hours — time enough to outlast the firestorm or any later emergency. Long-term fallout was the enduring threat to our survival. The cumulative mean lethal dose is 200 rads for an adult and much less for children. As built, the shelter would reduce their expected 30-day exposure to that associated with a dental X-ray.

Having done everything we could to build and stock the shelter for all contingencies, life settled back to normal and we enjoyed our family room for nearly two years before public attitudes and the international situation began to change. In the interim we showed our shelter to about 200 visitors and interested parties, including the New York State Director of Civil Defense.

The growing Soviet strategic threat finally began to penetrate public perception in mid-1960. Press coverage of the civil defense problem until then had been either minimal or negative. Then Jess Gorkin, the editor of *Parade* Magazine, whom I met during a flight in July 1961, agreed to do an article on our shelter in the public interest. Veteran reporter Sid Ross and his photographer came for a visit and wrote a very thorough and sympathetic article on our shelter efforts and rationale. This was included as an inset in a two-page article on "Where the World Leaders would Hide," with large pictures of Kennedy, Khrushchev, Macmillan, and Mao Tse Tung and descriptions of their shelters. The *Binghamton Press* did a more detailed piece that same Sunday edition of the Press.

During a two-week classified session at the USAF Air University at Maxwell Field in Montgomery, AL, in early 1961, I gave a fellow classmate, a lt. colonel from the Office of Civil Defense Management (or OCDM) in Battle Creek, MI, the shelter plans. The next issue of the OCDM *Shelter Bulletin* included some of its design features.

Negative international developments during 1961 warranted a big increase in defense spending and warnings to the Soviets regarding their threats to Berlin. They responded by starting to build the Berlin Wall that remained in place for 28 years.

Under Thomas J. Watson, Jr., IBM was sensitive to the worsening trend in US-Soviet relations. IBM was a small-town company. Most of IBM's manufacturing plants and laboratories were located in small urban areas. This was one factor in the decision to move half of the world headquarters staff from 590 Madison Avenue in New York 40 miles north to Armonk in mid-1961, "As a temporary experiment in non-urban corporate management location." They were followed by the remaining staff six months later. IBM Corporate Management has remained there for the past 40 years.

IBM established a Family Shelter Assistance and Provisions Program for all employees. The IBM-Owego house newspaper ran a major feature on our shelter in November 1961. Later, and for the first time, the IBM Board of Directors met in the Owego facility to stress the defense contributions of the Company.

When the Cuban Missile Crisis occurred in the fall of 1962, we simply refreshed the stored water and checked all of the shelter provisions. We had anticipated and prepared for such an event four years earlier. We all stayed close to home during that October.

On January 22, 1963, Corporate Headquarters extended an invitation from Tom Watson to discuss IBM Corporate civil defense policies. He actually wanted a critique of the plans for his existing family shelter in Pawling, NY. Watson made some significant changes based on my input; this visit provided a little more security to my sometimes tenuous position in IBM-Owego.

The family supported, or at least tolerated, our shelter venture ("There goes my fur coat") and provided appealing pictures for the newspapers. We tried hard to be constructive, both privately and publicly, on a very difficult and important national problem.

Chapter 3. The Space Guidance Center and Nuclear-Powered Cruise Missiles

Becoming part of the "Space Race" with the Soviet Union appealed to the higher management at both the IBM Corporate and Division levels. The Owego facility was renamed the Space Guidance Center. The Systems Research Department was renamed the Space Guidance and Control Department. The change reflected intent rather than reality.

Two engineering graduates and their associate professor were hired from a New England college to enhance the department's analytical capabilities, but they spent more time in highly theoretical discussions and lectures than in productive endeavors. The bright and vulnerable new professionals were an odd fit, and the SG & C department, where the atmosphere was competitive rather than cooperative, became even more stressful.

Many personnel changes had taken place. ESC growth required new, talented engineers. In 1956, Arthur DuBois joined IBM-Owego on the B-70 radar systems. Ed Smythe was hired that April. He later transferred to IBM-Huntsville, AL, to become IBM Chief Engineer on the Apollo Space Program and the later Space Shuttle Program. Dr. Howard Robbins and Dr. Dick Hillsley and Sherman Francisco were also much-needed, bright new hires. Years later, Sherman did the mathematical analysis of Einstein's Special Theory of Relativity that underlay the precise cesium clocks in the orbiting satellites of the Global Positioning System. John O. Cooney joined the B-70 BNMGS Program about this time, bringing practical expertise in radars, displays and lasers. He was also experienced in overall operational systems and simulations.

The Military Products Division marketing organization was largely unnecessary during the early years, but when the intrinsic vulnerability of the SAGE and B-70 programs became more apparent, Corporate and Division managements reacted by increasing the marketing staff. SGC-Owego Marketing was a traditionally overly-empowered group with frequent and influential access to senior management. Of course, my now-established independent, logically defensible, and proven analyses of our business areas were both an anathema and a threat to their status.

SGC-Owego was premised and structured to provide large avionics systems for vehicles that could maintain the strategic atmospheric threat to the Soviet Union. The B-70 was now an admitted failure; the B-58 could not replace the B-52; and the B-52 was not yet configured for survivable, low-altitude penetration or missions that could compete with the large and more effective ICBM and SLBM forces soon to be deployed.

Our purpose was to pursue and perform on programs that would have a valid and large business potential during the early 1960s. These included: (1) a radiation-resistant computer and guidance system for an intercontinental, nuclear-powered cruise missile; (2) re-instrumentation of the B-52 for low altitude penetration and real-time bomb damage assessment (BDA) for selective immediate re-strikes on hardened targets of the Soviet strategic counterforce, and (3) a new and more survivable supersonic heavy bomber to more effectively perform the new B-52 mission in the 1970s. Efforts to advocate and develop these programs and systems were concurrent and inter-related during the next two decades but are necessarily described separately.

The SGC marketing and management milieu made it an uphill battle all the way.

SLAM

The System Requirements Board review at Andrews AFB in mid-1957 included System Requirement No. 149, which described the vehicle and the mission for a new Strategic Low Altitude Missile (SLAM). This 40-foot cruise missile was to have a nuclear-powered ramjet engine capable of sustaining Mach 3 flight for 24 hours to deliver four one-megaton nuclear weapons to separate hardened targets from an altitude of 200 feet. This was to be the unmanned weapon system that could sustain our intercontinental atmospheric threat to the Soviet Union.

SLAM was to have the ugliest conceivable mission profile. The missile would be launched using a chemical booster to attain the airspeed of several hundred miles per hour needed to propel the missile off-shore and start the nuclear ram-jet engine. The reactor-powered ramjet would then accelerate the missile to Mach 3 for crossing the ocean at medium to higher altitudes.

After routing to avoid allied and neutral territories, the missile would descend to low altitude for penetration of the Soviet airspace. At an altitude of 200 feet and a speed that covered a mile in less than two seconds, SLAM would be essentially impervious to any then-foreseeable air defense systems.

The missile was to be programmed to accurately deliver the four nuclear weapons to separate targets. After that phase of the mission, SLAM would fly over most of the Soviet cities and other selected soft targets to destroy their buildings with the overpressure generated from the Mach cone during supersonic low altitude flight. After 20 or more hours of such catastrophic destruction, the missile would crash into an area selected for large-scale nuclear contamination from the nuclear reactor.

Building SLAM required many difficult technical developments. The AEC Lawrence Livermore Laboratory had spent several years secretly developing the Pluto reactor for the propulsion system and the Marquardt Corporation was designing the ramjet engine. The airframe, special steels and titanium structural materials were being researched by the Chance-Vought Aircraft Company (CVA), giving them a leading position to be the eventual prime contractor for the weapon system. Other possible missile prime contractors included Convair-San Diego and North American Aviation (NAA).

The SLAM guidance system would require an accurate inertial platform, a terrain-comparison (TERCOM) map-matching subsystem, and a radiation-resistant computer (RRC) to store and provide guidance data to the flight control system. The guidance system would have to function for 24 hours at high ambient temperatures and in a radiation environment that would kill a human in 15 seconds.

From December 1957 through June 1958, IBM conducted liaison with various Air Force Commands. Subsequently, a teaming agreement was arranged on the guidance and control system for BOLO, the NAA version of the SLAM Missile. The NAA Missile Division considered the IBM study to be the best of the dozen proposals submitted and separately sent our documents to the Air Force as part of their study contract. We next prepared a Development Planning Proposal for the BOLO Missile Guidance and Control System. Only Goodyear and IBM were selected from the dozen competitors.

The fundamental computer technology was based on pre-emptive work by Bill Maclay and Bob Betts. The system study team principals were Sherman Francisco, Howard Robbins and Ed Smythe. The first proposal was a literal response to the terms of the Request for Proposal (RFP).

We also submitted an unsolicited proposal that included Military Applied Research Program (MARP) funds to cover the several extensive supplemental studies necessary to define the Radiation-Resistant Computer (RRC) characteristics. The Air Force accepted our more ambitious proposal as a sole-source contract for all of the $180,000 in available funding. This step eliminated our competition for the SLAM Guidance and Control System. On June 3, the IBM-Owego general manager, Curt Johnson, called to congratulate us on winning the RRC Program.

I was named Engineering Coordinator (ad hoc project manager) for the RRC Program and was responsible to provide overall direction, objectives, scheduling and to allocate funds. The Owego functional organizations were to provide engineering and laboratory support. I also generated or approved all external reports to the Air Force. Eventually, IBM obtained all of the funds allocated by the Air Force for this type of development during 1961–1962, a total of $1,100,000.

The RRC was to be IBM's principal contribution to the SLAM Guidance and Control System. We proposed to solve the many associated problems by ambitious shielding, variously using lead and lithium hydroxide. To keep shielding to an acceptable minimum, Esaki diodes were selected as the most radiation-resistant semiconductors for the digital computer.

A small rotating magnetic drum memory was designed by Don Carter as the major data storage unit for the computer. Ferrite core matrices provided the memory needed in the computation process. Both memory devices would be invulnerable to the intense nuclear radiation environment. The computer was packaged in a toroid of individual wedge-shaped elements surrounding the cylindrical drum memory. The shielding also reduced the cooling problems from the high ambient temperatures generated by the Pluto reactor.

The main computer functions included system control of the reactor and ramjet engine, initial alignment and in-flight corrections for the inertial platform, continuous flight control signals, system control of the precision altimeter, and digital map-matching to determine SLAM present position and provide continuous course corrections to maintain the pre-programmed flight path.

Three computer engineers designed a serial processing one-channel computer capable of 70,000 operations per second. This capacity was the best then realizable within the constraints of radiation-resistant technologies and space available for the necessary SLAM functions. Forty years later, unrestricted 64-channel parallel processing can provide over a trillion operations per second.

Inertial platforms of the early 1960s could provide acceptable flight path accuracy of about one degree of drift per hour. Although SLAM would be able to cross the Atlantic Ocean at 2,000 miles per hour, more than one hour would elapse between launch and continental landfall. This required digital map-matching over off-shore islands such as the Azores, Crete and the extensive string of islands north of the Soviet Union.

TERCOM, or terrain-comparison map-matching techniques, were being developed in the late 1950s. Digital maps of the terrain altitude in areas around designated fix-points were to be prepared and stored on the RRC drum memory. SLAM would be programmed to maintain an absolute fixed altitude above the terrain and the precision altimeter would determine continuous altitudes under the SLAM flight path. The computer would compare the sensed altitude with the stored maps to determine the missile's present position and generate guidance signals needed to reach the next checkpoint. Thirty years later, comparable techniques were used to conduct the massive bomber, shipboard and submarine cruise missile launchings against Iraq during Desert Storm.

Most of the strategic targets in the Soviet Union were in the vast flat areas of the country. Topographic anomalies such as lakes, rivers and the occasional hill were used to improve the accuracy of the map-matching process.

Jerry Gwin, Dr. Robbins and I went to the Livermore Laboratory to obtain the data needed to control the reactor with the central RRC. They centered the meeting on a 20-foot Pluto schematic that was very detailed yet lacked the specific data that we sought. This technical "snow-job" continued until our Dr. Robbins pointed out that the schematic had an open loop in the hydraulic servomechanism controlling one of the scram rods. Thereafter, we enjoyed their full cooperation.

Irving Hoffman was the Pluto program manager at the Atomic Energy Commission Headquarters at Germantown, MD. We developed a mutually useful professional friendship that lasted for several years. Early on, he and his staff were given the "Environmental Basis for Strategic Weapon System Development" briefing on why IBM was energetically participating in the SLAM/Pluto Program.

In the early 1960s, ICBM and SLBM technology and deployments promised enough missiles and warheads (especially with the evolution of multiple re-entry vehicles — MRVs, and later multiple, independently targetable re-entry vehicles — MIRVs) that these weapons could be used to prepare penetration paths for the strategic bomber force. This would greatly improve the survivability and effectiveness of the bombers, enhancing their credibility as a low-altitude strategic deterrent.

Congress had perceived the foregoing trends and, reasonably, was questioning the need for the SLAM/Pluto Weapon System. Whatever the program's eventual merits, the high cost, exotic technologies, operational risks associated with nuclear propulsion, and the growing bomber force were factors that diminished congressional support for the SLAM Program.

Irving Hoffman asked that I testify to Congress, using the rationale presented to him a few years earlier. IBM did not want to take positions on such issues, but the AEC insisted. We later briefed some of the AEC commissioners and assistant general managers at Germantown, MD, for their approvals.

On April 19, 1962, accompanied by Irving Hoffman, Clint Grace as the very reluctant SGC manager of engineering, and a coterie of lesser managers and marketeers, I was sent to brief the staff and several members of the Joint Congressional Committee on Atomic Energy. In the early evening, we went into the Capitol Building through a ground-level entrance to their second floor conference room underneath the Capitol's front steps. Following a preliminary statement by Clint Grace to put as much distance as possible between IBM and this event, the three-hour briefing, testimony and discussion were well received. However, the occasion remained very unpopular with SGC management.

Pluto Reactor

As a result of the foregoing and other lobbying efforts, the AEC obtained funding sufficient to demonstrate the Pluto Reactor independent of the SLAM Program. The latter was in foreseeable decline, a redundant and expensive solution to the deterrence problem that was being solved by the evolving Triad of bomber and ballistic missile forces, it faded away in the mid-1960s.

Two years later, Irving Hoffman extended an invitation to Nevada to witness the first and last full-power test demonstration of the Pluto Reactor. At the Jackass Flats Test Site, a huge assembly of very high pressure cylinders covered a nearby hillside. This facility would provide the Mach 3 air supply necessary to simulate two or three minutes of SLAM flight. Without this huge rush of high pressure air through the reactor core, it would immediately and violently self-destruct.

The prototype test reactor was mounted on a full-sized flat-bed railroad car and moved along a 400-yard track by a small railroad switch engine. The test was controlled from a bunker 40 feet underground and several hundred yards away from the reactor. During the two-plus minutes of reactor operation, the bunker shook so violently that it was hard to follow the TV monitors mounted on and around the flat car.

The noise and ground vibration was so intense that we had to shout to be heard. Anyway, nearly everybody was speechless. During that brief interval, mankind generated more non-explosive power from a single small source than by any other event in history. There were no press notices.

Following the test, and before we could come out of the control room, the unmanned switch engine moved the flat car into a huge cut-and-cover concrete and earthen bunker with steel doors over two feet in thickness. The doors were later welded shut and covered with 20 feet of earth. Given the expected rate of decay in radioactive emissions, the reactor crypt must remain sealed for several hundred years.

Radiation-Resistant Computer

Due to the failure of the functional managers to perform during 1961 and 1962, there were extensive restructurings of the RRC Program. Continuing failure to perform on the revised program caused the engineering coordinator to resign that position on March 21, 1962.

In February 1963, IBM was asked to perform an avionics study for the Multipurpose Long-Endurance (MPLE) aircraft proposed as a new strategic bomber. The RRC and supplementary system studies were used as the guidance for the attendant Conventional Low Altitude Missile (CLAM) guidance system. With Strategic Air Command (SAC) encouragement, this study was presented to ASD-Dayton, AFSC and USAF Headquarters.

In March, we negotiated with the NAA Missiles and Space Division to arrange a joint study effort on CLAM beginning on April 15, 1963. The arrangement was sought by NAA due to the endorsement of IBM's MPLE work and our prior association with them on their BOLO Program in 1958–1959.

The RRC design engineers set up the bench model of the RRC and programmed it for representative missions. The RRC ran 24 hours a day for a year without a glitch. This was quite a technical accomplishment in 1963–1964. Then they turned the RRC off and put it in the junk pile.

Chapter 4. Strategic Bombers, Missiles & Armaments

The cancellation of the B-70 caused professional and organizational trauma, created confusion and aroused severe self-interests among all involved, including IBM-Owego marketing and the program and engineering organizations for the B-52 and B-70 bombing systems.

My analysis of the B-52 and B-70 future operational environments helped the Air Force to formulate an Advanced Development Objective (ADO-22) to study the kinds of missions and the relevant avionics requirements that could be envisioned for armed reconnaissance by heavy bomber aircraft. Art Cooper accompanied me to ASD-Dayton on May 19, 1961, to brief them in support of ADO-22 and to further justify a major strategic bomber avionics development within that framework.

Three weeks later, I briefed the FSD President to obtain three man-years of funding to pursue ADO-22. After years of unsuccessful efforts to remain competitive in the strategic avionics market, the current IBM-Owego product lines would be obsolescent within three years.

Funding was granted with the understanding that I would conduct the study as the Engineering Coordinator, reporting to the manager of engineering. I produced a detailed paper, "B-52 Environment & Utility," that included bombing mission profiles, ground rules, and the responsible departments.

During May–November 1961, many Air Force offices were briefed to test and refine the direction given to our ADO-22 Study Group; this included ASD Plans, Offensive Systems Branch, Task Group 9 (General LeMay's study group), Foreign Technology Division, and the B-52 SPO. Encouraged, we briefed the USAF HQ Development Planning organization and their Guidance and Control Board. Still later, we visited SAC HQ Requirements and their DCS for Intelligence. The paper was also given to the DOD Weapons System Evaluation Group and the RAND and ANSER corporations.

In June 1961, Curt Johnson retired as IBM-Owego general manager; he was replaced by Art Cooper. Clint Grace became the new engineering manager. Clint had been among the early managers of the B-52 and B-70 programs and relied on a network of former associates. The functional and project managers under the new boss had no interest in the integrated avionics system study represented to the New Business Review Board.

In early July, Grace directed me to share the study with another B-52 program office effort to define improvements in the current IBM bombing system. By October 1961, the "establishment" managers had deferred the ADO-22 Study by five months. They benefited substantially from our ground rules but the B-52 effort was of minimal impact and the B-70 work went nowhere. Over half of the study funds were expended on their "B-52 Flexibility Study."

Nothing was being done on the ADO-22 commitment. I tried to raise an alarm with Clint Grace on October 16 but found him unresponsive.

FSD Headquarters reviewed of all IBM-funded reconnaissance–strike pre-proposal efforts on November 21. The Owego B-52 departments had completed their "B-52 Flexibility Study" and were confident it met both SAC's requirements and IBM's business interests. To them, the question was whether to continue the B-52/ADO-22 Study. The SAC Requirements Office disagreed; they found the IBM Flexibility Study an inadequate response to the foreseeable needs of the SAC B-52 forces and encouraged the ADO-22 work to go forward.

B-52 Re-Instrumentation Study

We met with the NBRB on December 4, 1961, to request re-initiation of the B-52 Study. I had to insist on adequate staffing so that I could reassure the New Business Review Board (NBRB) on January 12 that we would perform the B-52 Re-instrumentation Study as currently scheduled — otherwise, I said, I would have had to recommend that they cancel the ADO-22 Study. Using Division Management to manage your supervisor is not the best way to get ahead in IBM (or anywhere), but sometimes it achieves the immediate objective.

Known for five years of effective and responsible liaison with DOD and the major Air Force commands, I was asked by the DOD Weapon System Evaluation Group for briefings on all of the IBM reconnaissance–strike studies on the B-52 and B-70. Separate IBM organizational briefings were negotiated for the B-52 and B-70 organizations. I screened their proposed presentations for errors and inconsistencies. Our intentions for the now-current B-52 Re-Instrumentation Study were presented separately.

Now that Art Cooper was general manager at IBM-Owego, he was advised that a certain IBM senior engineer with a national reputation in a narrow field of endeavor had never had an opportunity for a line management position. Art scheduled me to attend the IBM Management School at Sands Point, Long Island, in April. Allegedly an intense course of study, this was a great six-week vacation compared to the past seven years.

Returning from Sands Point, I learned that the B-52 Re-Instrumentation Study had been re-directed for completion by May 21 without an adequate definition of the air-to-surface missile guidance subsystem, missile/bomber interface, or the new B-52 Central Digital Computer. Since these equipments were to be the primary IBM hardware, software, competitive advantage, and revenue sources, I extended the study for six weeks to complete the scope represented to NBRB in May and November 1961.

Due to the TFX Proposal (avionics system for the F-111), unavailability of critical personnel, and impending transfer of two principal systems engineers to the new

NASA/IBM facility in Huntsville, AL, the study progressed slowly. August 1962 had arrived with little hope of completing it.

This time I called SAC Headquarters to ask them to prod the recalcitrant IBM management. Colonel Pat Montoya called Grace and/or Cooper and asked, "Where is that B-52/ADO-22 Study that you promised us several months ago?" The results were immediate and constructive. Montoya followed up in early September, while I was re-engineering and writing the summary of the entire B-52 Re-Instrumentation Study. A representative of the SAC Requirements Office visited Owego; after a brief review, he asked for a presentation to SAC during the first week in October. In the end, the four-month study had required 16 months for completion.

Arthur DuBois defined the TV, infrared and radar subsystems. Vouter Vanderkulk did the applied mathematics required to locate the Soviet anti-aircraft radars by range-on-bearings-only data processing. Sherm Francisco solved the in-flight alignment problems associated with the newly-defined small ASM (SRAM). Our final report described a new BNMGS and Small ASM that would allow the re-instrumented B-52 to prevail, alone, if necessary, against the Eurasian air defense system and target structure of the late 1960s.

The B-52 Re-Instrumentation Study was briefed to about 30 field grade officers at SAC Headquarters on October 3. They endorsed our work for another presentation that afternoon to Major General Russell and Brigadier General Crum of SAC Operations and Plans, respectively. The generals encouraged an immediate presentation of our study to ASD-Dayton senior management under the cognizance of the ASD B-52 SPO. The subsequent briefings were delayed for five weeks due to the Cuban Missile Crisis.

During November 1962 through March 8, 1963, the B-52 Re-Instrumentation Study was briefed throughout the Air Force. The demands of the bomb damage assessment and re-strike mission, and IBM solutions to these problems, were presented again at SAC, ASD, OCAMA, AFSC and USAF Headquarters: some 200 senior officers and civilians representing 30 offices in all.

Generally, the Air Force response to our briefings was one of endorsement, in principle; recognition of ultimate utility; confirmation of technical realizability; — and severe reservations regarding the cost of the entire B-52 re-instrumentation. The DOD and Air Force budgets were then stressed by the most expensive phase of the Minuteman ICBM deployments — a much more important contribution to America's strategic force posture.

We had provided the Air Force a plan for reacting to their strategic bomber problem, exercized initiative, and been accredited with responsible independent intellectual significance in the BNMGS Industry. This protected and furthered our business interests in the primary IBM-Owego product line.

Air Force Priorities

Accepting the validity of the IBM/ADO-22 Study, the Air Force sequentially prioritized their responses. The small ASM was a good way to extend the useful life of the B-52 under any circumstances; the electro-optical sensors proposed for low-altitude bomb damage assessment could have immediate peripheral utility; and the eventual mission of real-time bomb damage assessment and immediate, selective re-strike against hardened targets could be implemented when required as a "force multiplier" in later years.

The Small Air-to-Surface Missile, or SRAM

USAF HQ used our final report to define the operational parameters, characteristics and flight profiles of the new Short Range Attack Missile (SRAM). The SRAM air-to-surface missile was to be capable of launching at both high and low altitudes from the B-52 wing pylons or internal rotating multi-missile launchers. Due to the expected moderate accuracy, the ASMs initially would be assigned to soft-area targets such as Soviet air defense missile batteries and interceptor airbases. A nuclear warhead with a yield ten times greater than the Hiroshima bomb would assure a high probability of destroying these targets at distances up to 100 miles.

The Air Force eventually deployed more than 1,000 of these missiles on the B-52 and later FB-111 and B–1 bombers. Thus my five-year initiative to conceptualize and further the SRAM Program extended and enhanced the effectiveness of the B-52 in the strategic bomber mission for another 20 years.

Electro-Optical Sensors

Advances in television and infrared scanning devices allowed Arthur DuBois to configure a very capable TV camera and a forward-looking infrared (FLIR) sensor that would display high-resolution pictures at short ranges. These sensors were essential for the BDA/re-strike mission but also offered an immediate and enduring utility of great interest to SAC.

SAC airbases are designed for the rapid sequential take-offs of all bombers and colocated refueling tanker aircraft to avoid a Soviet SLBM and ICBM barrage that would destroy the base and all aircraft on the ground or in nearby flight paths.

B-52 cockpits contain shielding curtains to protect the pilots from flash blindness due to nuclear bursts that do not destroy the aircraft. These curtains were normally closed in varying degrees during emergency taxiing and take-offs. The electro-optical sensors, when mounted on the late-model B-52 G & H models, permitted the pilots to do rapid "blind" taxiing and take-offs with greater safety from collisions or veering off the concrete.

For decades after World War II, the principal methods of high altitude bombing using airborne radar were direct target sighting and using off-set aim points. The latter technique was used when the target afforded a poor radar image. The off-set aim point, selected for good radar returns, and the computed bomb release point permitted the bomber to maintain a course over the less discernible target. The weapon delivery would then occur regardless of possible prior attacks by missiles or aircraft.

That mission would be stressful enough for any bomber crew, but the operational environment was foreseeably and rapidly changing for the worse. The Soviets were beginning to deploy very effective long-range surface-to-air missiles (SAMs) with small nuclear warheads. Their first such missile to be deployed, the SA-5, similar to our Nike Hercules, would be an effective defense against the B-70. The B-70 was already potentially defunct; but the Soviets' peripheral and interior deployments of the SA-5 and other smaller SAMs would also be deadly if used against the B-52 at high altitudes. These and later SAMs would simply preclude effective high-altitude penetration and weapon delivery by strategic bombers. Low was the only way to go.

Real-time BDA and re-strikes against hardened targets or targets of imprecisely-known locations would eventually become the principal unique and necessary mission of the manned strategic bomber. The rapidly increasing numbers of more effective ICBMs and SLBMs would eliminate the need for less cost-effective bomber operations against soft military targets and urban/industrial complexes. This trend also reduced the need for large numbers of strategic bombers.

As the weapon delivery accuracy improved and numbers of ballistic missile warheads increased through the evolution of maneuvering, independently targetable re-entry vehicles (MIRVs), first-strike weapon delivery missions by bomber aircraft against hardened targets would no longer be needed. The bomber missions would be greatly reduced in numbers and would become primarily BDA and re-strike against only hardened targets. Two MIRVs against a soft target would produce such a low probability of survival that prompt BDA by bombers in the first days of a general nuclear war would be a waste of resources.

The Air Force now acknowledged these operational trends. Low altitude target sighting using the bomber's new high resolution sensors was first proven by flying low over US ICBM silos. With the silo doors open, the bombardier/navigator could readily see the nose cone of the Titan ICBM.

During the 1962 re-instrumentation study, we devised a method to determine the probability of kill on a given target by measuring the distance from the crater created by the prior attack. Since the bombardier would know the target hardness and yield of the previously delivered weapon, automatic computations would determine whether the target had been destroyed or required an immediate re-strike by the bomber. If so, this could be done using a drogued bomb or an SRAM fired over-the-shoulder from the B-52. The latter would be less effective, but two overflights on a single target during the same mission was an unacceptable alternative.

The greater difficulty of this mission for the aircrew and the gradual evolution of the need delayed potential operational utilization for many years. However, once in the mid-1970s I was having lunch at SAC Headquarters with an Air Force colonel and casually mentioned the BDA/re-strike mission; I was immediately cautioned not to speak of it in the cafeteria. This implied that the BDA/re-strike mission was now part of the war plan. I later referred the colonel to the Congressional Record, in which the Air Force in the late 1960s used one of my unsolicited papers (without attribution) to justify the B-1 Bomber to skeptical senators. The mission, first conceptualized in 1957 and brought to fruition with the 1962 B-52 Re-Instrumentation Study, had finally been realized 16 years later.

Multipurpose Long Endurance Aircraft

The Multipurpose Long Endurance Aircraft (MPLE) was an Air Force concept for a very large subsonic aircraft that could be configured to perform several different missions. These included stand-off long-range cruise missile (CLAM) launching without penetrating the very daunting Soviet air defenses. Other envisioned missions were the Airborne Command Post, troop and materiel transport, aerial tanker support, and possible anti-submarine patrol operations. However, even if it were successful, the aircraft would necessarily be sub-optimum for each of these missions due to high cost and/or lesser capabilities of specialized aircraft such as the B-52, C-124, KC-135 and the Navy P2V respectively.

On February 15, 1963, the SAC Requirements Office requested that IBM, as the only avionics manufacturer, help them in preparing a Special Operational Requirement (SOR) for the MPLE by defining the instrumentation for the new vehicle and associated missiles. We scrambled to meet the SAC request and briefed them on February 26.

Still, I had fundamental concerns about the validity of the MPLE concept. The SAC staff was separately advised that the briefing was only a literal response to their requirements and that several serious questions needed to be answered before the intrinsic military validity of MPLE could be established. SAC asked that IBM brief ASD, AFSC and USAF Headquarters March 6–21, 1963.

On March 14, the Air Force recommended IBM to assist North American Aviation (NAA-Los Angeles) in their CLAM study for MPLE. FSD effected a potential teaming arrangement with NAA in Washington, DC, and Los Angeles. NAA wrote the IBM affiliation into their missile proposal.

IBM Diversification

Meanwhile, throughout the early 1960s the IBM-Owego marketing people badgered our technical people for proposal preparations on numerous concepts and programs for which we were not competitive, or that implied unrealizable costs, lacked military validity, or could not provide the potential revenues needed to support our large facility. Such ventures included the Minuteman II ICBM Guidance and Control, Space Plane, TFX, Dynasoar, ANP (a nuclear-propelled bomber), Advanced Guidance Program (AGP) and others.

The Dynasoar Program (dynamic soarer) was pursued by the Air Force and industry for several years. The proposed aerospace craft, about the size and shape of the small A-4 fighter-bomber, was to be rocket-boosted into near space to a high supersonic speed. The spacecraft would then glide into a shallow re-entry flight path and bounce off the denser air to dynamically soar back into space, skipping over the Soviet Union for manned reconnaissance and possible weapon delivery. I visited Boeing-Seattle twice for very non-productive sessions with their Dynasoar program managers.

Consequences of the B-52 Study

The effort to help ensure a timely and intellectually responsible evolution of strategic bombing missions and the attendant instrumentation was rapidly coming to fruition in 1963. The "B-52 Re-Instrumentation Study" began to attract unsolicited interest within the industry. From June 1962 through February 1963, Boeing repeatedly and unsuccessfully requested that IBM Management and the Air Force provide them information regarding the content and progress of the study.

Boeing Company

On February 15, Boeing-Wichita management visited IBM-Owego to discuss a common business interest in the B-52. Following an exchange of correspondence for proprietary agreements, Boeing-Wichita was briefed on April 5 regarding the B-52 Final Report. The effort was well received and subsequent contacts by Boeing encouraged a possible IBM-Boeing teaming for market development.

Martin Marietta Company

At the suggestion of the Air Force, Martin-Orlando contacted me to explore the possibility of teaming to develop and produce the SRAM and attendant modification of the B-52 BNS. Ken Driessen, now in Owego Marketing, accompanied me on a visit to Martin-Orlando to provide the basis for a proprietary study and joint marketing endeavor. We delineated the tasks to be performed by each company on March 13.

On March 27, an IBM study team went to Orlando for the exchange of proprietary data and to finalize a working agreement. I briefed them on both our B-52 and

MPLE studies. After that, the meeting did not go well. The Martin team isolated me from the IBM technical and contracts groups in an attempt to drag more proprietary information out of them while I was otherwise engaged. The Martin people were insisting that IBM would not design and manufacture the ASM guidance system, our principal potential revenue source from any later SRAM Program. In spite of Arthur DuBois' effort to hold the line on data exchange, things got out of hand, to the point that I finally advised them that we had accomplished our initial work on teaming and that we would be pleased to see them in Owego when they had resolved their position on teaming. And we left.

The Martin team, well aware of my contributions to the Air Force, knew they could more easily control a Martin-IBM team without my involvement. IBM-Owego management and marketing privately shared these views — now that the program and IBM's position in it were well established, they wanted more than ever to have sole control the business prospects.

When we returned to Owego, Clint Grace stated that due to my "very serious disruption" during the visit, there would no longer be an Engineering Coordinator on the B-52/SRAM Program.

By April 9, I was an assistant to the manager of the Engineering Laboratory. We established a teaming arrangement that retained the SRAM guidance business for IBM, evaluated Martin's teaming proposal vis-à-vis our ground rules, arranged a data exchange, and established a task delineation agreement.

During the next few months, Martin euchred IBM-Owego entirely out of the SRAM guidance subsystem. Later, the Martin-IBM team lost the entire SRAM Program to Boeing.

On April 24, a headhunter called me at home to offer a job on the SRAM Program in Martin-Orlando, in spite of my "disruptive behavior" there on March 27.

New Business Plan

Cooper and Grace directed me to contact USAF and DOD to formulate a new business plan. In December 1962 and mid-March 1963, I called at ASD, AFSC and USAF Headquarters, and DOD Research and Engineering; I recommended four high-priority business activities:

1. On the RRC/SLAM/CLAM, we should team with NAA as had tentatively been arranged. We should also try to apply the RRC/SLAM technology to the Navy's Advanced Sea-Based Deterrent Program.
2. FSD should team with Martin on the B-52/SRAM to provide the SRAM Guidance System and the attendant degree of B-52 BNMGS re-instrumentation. This was the most promising way to derive new funds from the B-52 study and develop new, detailed cost estimates.
3. For MPLE/ERSA/RS-52, we should submit an unsolicited proposal to AFSC HQ for a generalized BNMGS study applicable to this type of aircraft. Our existing B-52 and MPLE studies were an adequate basis for this task. (I personally favored the Extended Range Strategic Aircraft, ERSA, as the next heavy bomber development, but I had not yet discussed that analysis with anyone.)
4. Current obligations on the B-70 Program should be fulfilled, producing two prototype BNMGS and flight testing one of them. There was little hope for a significant B-70 business potential.

For seven years, overcoming severe and enduring internal management opposition, I had provided the intellectual basis and much of the initiative for the perception, re-direction and furtherance of IBM's primary business interest in strategic bomber avionics. During late 1962 and early 1963, the professional environment had become too demanding for any single individual. Fortunately, we had prevailed in the RRC Program and all of the major new business prospects facing IBM-Owego. However, these years of exceptional effort could no longer be sustained.

Thus when Cooper and Grace, likely with their plans already agreed upon, asked how to pursue these newly established business opportunities, I was ready. First, I suggested, Grace should continue the RRC/SLAM/CLAM efforts under Fred Foss in Advanced Systems Research. Then he should form a new department, under my management, reporting to him, to further develop our business interests in future bomber programs. These would include B-52/SRAM, B-52 Re-Instrumentation, and the later heavy bomber developments such as ERSA and MPLE.

I had developed our positions in the foregoing programs from a staff position, by default, for the last several years. A new effort was needed because of the enduring inability of the established B-52 and B-70 organizations to protect and further their business interests. A small number of the lead professionals who had contributed to the B-52 and SRAM studies would form the cadre of the new department. The established B-52 organization could continue to perform under current contracts and future extensions to develop and manufacture the operational B-52 BNS. They would later accept responsibility for any large new programs developed by the proposed entrepreneurial department for the B-52/SRAM or later heavy bombers.

The B-70 organization would complete the contractual development of two prototype BNMGS and the flight test. Their eventual termination would be eased by gradual transfer of personnel to the B-52 Program and the department being formed.

On April 1, 1963, Clint Grace advised me that he was transferring all new business program responsibilities to the three established business organizations, and he suggested I stay on at Owego as his technical advisor. I gave him a prepared letter requesting a transfer to FSD Headquarters. During the next three years, IBM-Owego realized absolutely nothing from all the initiatives and the business opportunities established during 1960–1963.

A New Strategic Bomber

SGC-Owego was generally a peripheral business to IBM and was rather self-contained. The newly formed Federal System Division Headquarters, recently relocated to Rockville, MD, had few people who understood the military aerospace avionics business. A recently transferred senior engineer with far too much Owego experience was a welcome anomaly. The new assignment was the first time in 13 years that I was spared an overload of urgent demands. I now had the opportunity and responsibility to quantify probable future revenues the major programs in the Division.

John B. Jackson was then manager of the headquarters staff. Dr. William F. Offutt, a former advisor to an Air Force general in the DOD Weapons System Evaluation Group (WESG), was my new boss and was a good friend for the next 40 years.

I projected that BNS business revenues would fall to three percent of that needed to sustain the Owego facility within three years. Similarly, FSD-Kingston's SAGE income and new business diversification would drop from $130,000,000 to about $40,000,000. The first two months were spent educating the staff and senior managers on the nature and decline of the Division's business prospects.

Diversions occurred when the military services asked SGC-Owego for briefings that they could not, as yet, provide. Accordingly, they merely tagged along on my presentations to the Air Force Space Systems Division's Project Forecast General War Panel in July and the Navy's Advanced Sea-Based Deterrent Study Group in September 1963.

Manned Aircraft Strategic Systems Group (MASSG)

The struggle to define a new strategic aircraft continued within the Air Force, DOD and the industrial community. The Manned Aircraft Strategic Systems Group (MASSG) within the Air Staff was conducting the necessary studies, assisted by the RAND Corporation and major aircraft manufacturers. Some of the principals included Colonels Wadsworth, Bottomly and D.C. Jones (mentioned previously with reference to the B-70), Lt. Colonel John Daily, and Majors Gilbert and Wally Hynds. They were trying to balance or comply with the many and diverse demands of senior officials that included General LeMay and Secretary of Defense McNamara.

On August 9, I heard from John Dailey, who was responsible for the $16 million contract that the Air Force afforded RAND for studies and analytical support. His concern was that RAND was not being effective on this problem. Having incurred General LeMay's considerable wrath by not earlier supporting his preference for the B-70, RAND had, in Dailey's judgment, become too political and indecisive.

The RAND approach was to consider 125 alternatives and variants through computer analysis and the newly popular concept of cost effectiveness. Simultaneously considering all aspects of the many bomber alternatives had simply swamped the RAND staff and their roomful of IBM 650 computers.

Dailey and others on the Air Staff, aware of my prior work on the B-70, B-52 Re-Instrumentation, MPLE and strategic force structure analysis, wanted me to examine the entire problem.

Bill Offutt and I spent a day selling the proposed arrangement with the Air Force to FSD Management. Don Spaulding was now the FSD President, which helped. Their approval was contingent on an Air Force understanding that the opinions provided were not those of IBM.

The next month I burned the midnight oil working on the unclassified aspects of the study. I took the qualitative analytical approach of making eight general categories of the 125 alternatives, then gradually and specifically generating the rationales to reject seven of the categories.

The remaining prospective weapon systems were necessarily large aircraft capable of carrying heavy payloads over intercontinental ranges and penetrating alone deep into the USSR or Peoples Republic of China. The principal alternatives were the Re-Instrumented B-52, a Multipurpose Long-Endurance Aircraft (MPLE), the Advanced Manned Penetrator (AMP), and the ERSA or Extended Range Strategic Aircraft. The latter two were similar concepts, but the ERSA could provide superior performance over a broader range of capabilities. SLAM was included as an adjunct, specialized weapon system. Several kinds of cruise missiles and SRAM variants were also considered as armaments for the large aircraft.

Ed Oliver, manager of the RAND Washington office, had written a preliminary summary paper for the Air Staff that supported a relatively small supersonic aircraft as the new strategic bomber. This was politically appropriate since Secretary of Defense McNamara had been reviewing supersonic propulsion for President Kennedy in relation to a United States supersonic transport development. McNamara also had some major engineering problems on the new TFX Program (later F-111), a large fighter-bomber designed for tactical interdiction missions a few hundred miles behind a battlefront.

The RAND paper proposed and McNamara supported an enlarged version of the TFX as the next strategic bomber. This would presumably provide the funds for re-engineering the TFX and a new strategic bomber. Further, tactical-range aircraft of this genre were consistent with the then-popular Herman Kahn and McNamara theories on "Limited Nuclear War" vis-à-vis general nuclear war.

McNamara subsequently developed the FB-111 as a separate but related program to realize a new manned strategic aircraft. Funding for this program and the on-going ballistic missile deployment doubtless delayed the process of developing a more useful new heavy strategic bomber — as it did for B-52 re-instrumentation. At least two wings of FB-111s were eventually deployed to the northernmost SAC bases, due to the severe range limitations of the aircraft. The FB-111s were also deployed to England on a rotational basis for the same reason.

In his testimony to Congress, McNamara noted that all of the associated development costs "were in his head." He also stated that the FB-111 could fly intercontinental distances, carry six SRAMs on wing pylons, and fly supersonically above Mach 2. He failed to add that the FB-111 could not do any two of these things at the same time.

The Air Force scheduled the presentation of the new paper on September 11. The Washington RAND people were very confident of their well-established position with the Air Force and their large ongoing, politically-atuned analysis. They considered me a newcomer to be politely ignored.

The meeting format resembled a friendly inquisition. Three RAND people, Ed Oliver, Phil Barham and another aeronautical engineer, sat on the opposite side of the table. They were surrounded by six colonels and another six or eight lt. colonels and majors, including John Dailey. The colonels and RAND people all had copies of my new paper.

RAND started with questions intended to support their position paper; but with some help from the colonels, I redirected the meeting to a detailed discussion and defense of the new study. RAND was rather taken aback by the Air Staff's interest in and support of these arguments.

When the RAND people tried to disprove the long-established theses on future strategic manned bomber missions that were fundamental to the paper, their arguments were readily defeated. The meeting lasted two hours. Some members of MASSG characterized the RAND position paper as "accommodative sophistry." They promised a thorough consideration of my paper.

John Dailey asked me to make a return visit to the Pentagon a few days later. He had taken considerable professional risk by inviting an outsider into the latest manned bomber mêlée and was inordinately pleased with the outcome. The Air Staff had rejected the RAND position paper and their massive approach to the analytical problem.

Lt. Colonel Dailey was quite amused by what he described as "having the RAND Corporation and their computer installation as a front for a guy in the back room with a slide-rule who decided the outcome." Ed Oliver soon transferred back to Santa Monica. Some months later, John showed me the final summary of the RAND effort, endorsing the BDA and Re-strike Mission and a new heavy bomber such as ERSA. They neglected to give any accreditation for their new insights into the future of the manned strategic bomber.

MASSG suggested several improvements to the paper, including a broader title, which was changed to "Strategic Aircraft Weapon Systems Environment and Utility" or the SAWSEU Memorandum.

MASSG agreed to control non-IBM distribution of the document due to the proprietary, broad and sensitive content, and my very informal and unacknowledged arrangement with the Air Staff. Eventually they distributed the paper to all concerned Air Force Commands, DOD and the Atomic Energy Commission.

The successful confrontation with the RAND Corporation helped to resolve the future evolution of strategic bombers. That would be a major enhancement to US strategic defenses and the first step in what turned out to be my 21 years of advocacy for what became the B-1 Bomber.

IBM-Owego management had heard rumors of my work with the Air Staff on the bomber problem, but they were not aware of the depth or outcome of the effort. Headquarters senior management had long understood that FSD would be provided with a proprietary summary that would be helpful to IBM-Owego business prospects and preparations, so after completing the Air Force work, I wrote up a summary, "Interpretation of Avionics Business Potential in Future Strategic Weapon Systems," with three "going out of business" charts that summarized their future revenues *sans* any new major programs.

I also identified likely future needs that Owego could aim to fulfill: the functions and new technologies for avionics developments related to new weapon systems. There were five evolutionary and progressive phases for the B-52: (1) defense suppression; (2) selective defense suppression; (3) BDA and re-strike; (4) complex BDA and reconnaissance; and (5) a completely integrated BNMGS for an RS-52. The avionics developments needed for each progressive phase could be related to the other simultaneous or later aircraft that included, in priority: (a) AMP/ERSA; (b) MPLE in four mission configurations; (c) CLAM and SLAM; and (d) naval weapon systems. High performance reconnaissance vehicles were also phased in, but not identified (SR-71). The final, secret SAWSEU Memorandum to the Air Staff was also included as IBMCD No. 20–100-386.

IBM Corporate Management then sent a new manager to head the headquarters technical staff. Don Spaulding replaced Charlie Benton as president and John Jackson was moved to SGC-Owego as assistant general manager. Instead of exploiting all the work Owego and Washington had put into the B-52, SGC-Owego set out to "diversify" the business. The bleak business future and excessive marketing organization developed a two-year business plan that included endeavors in tactical aircraft (logically), Advanced Sea-Based Deterrent (little capability), oceanography and supersonic transport (which would have represented wholly new fields of endeavor), and civilian (Apollo) and military space programs.

One day during this period, I was having lunch with Irving Hoffman at the Bethesda Country Club. He was still running the Pluto Reactor Program for SLAM and some other projects at the AEC HQ in Germantown, MD. We were interrupted by the announcement that President Kennedy had just been shot. It was November 19. Irving immediately departed to an undisclosed location, where he stayed for several days. He was one of the thousand or more senior government officials required to do this when advised of a national emergency.

Air Staff & SPO Support

The outcome of the confrontation with the RAND Corporation on September 11, 1963, was closely held by all concerned. The Air Force (and RAND) needed time to put the best face on such an abrupt change from political expediency to emphasis on the AMPS/ERSA concept for a new heavy bomber. The delay was particularly necessary to convince senior officials in DOD and the Air Force that AMPS/ERSA development was the best course of action. Eventually, the Air Staff asked me to present the SAWSEU briefing to more than 30 senior managers and technologists from other major commands, particularly the Aeronautical Systems Division (ASD) in Dayton, OH. There were no other IBM attendees.

Because of their interest in SLAM and their Pluto Reactor Project, the Atomic Energy Commission Headquarters requested the SAWSEU briefing on March 11, particularly seeking assurance on how SLAM could fit into the evolving strategic force structure. The AEC subsequently arranged for a briefing to the Department of Defense Research and Engineering (DDR&E) on March 19. DDR&E, particularly members of their Science Advisory Board, were more interested in the rationale for a new heavy bomber than the SLAM/Pluto programs.

The Air Force again renamed their new bomber concept, this time to AMPSS or Advanced Manned Penetrator Strategic System. A System Project Office (SPO) was established at ASD-Dayton with Lt. Colonel P.M. Spurrier as the first SPO Chief. The AMPSS designation was retained throughout the conceptual study phase and was finally changed to AMSA, or Advanced Manned Strategic Aircraft, for the program definition Phase. All of the major Air Force Commands now believed that the aircraft would be developed and eventually reach operational status — but we did not imagine that it would take 21 years!

Lt. Col. Spurrier requested the briefing for his new SPO on March 14. After careful consideration, he agreed with the mission definition and justification offered in the SAWSEU paper. The AMPSS SPO prepared a Preliminary Technical Development Plan, requesting a waiver of the new McNamara DOD Directive 3200.9, which would delay the AMPSS Program by a year or more. The DDR&E Science Advisory Board supported AMPSS in opposition to McNamara, lessening his resistance to the program.

The long-lead-time items in the development of a new strategic bomber were the engines and avionics. The initial $50 million allocated to the new AMSA SPO was divided into $45 million for the engines and $5 million for avionics studies. DOD Directive 3200.9 and other factors delayed the AMSA Avionics Study requests for proposals from July 1964 to May 1965. The engine development contract was awarded to GE-Cincinnati somewhat earlier.

Colonel Gilbert, from the MASSG, was the AMPSS Action Officer for the Air Staff. Based on my prior work with Lt. Col. Dailey, he asked me to go ahead and further define and justify the AMPSS Program. There were problems related to (1) bomber penetration using SRAM for defense suppression and (2) quantification of the BDA and re-strike missions over time in response to ICBM and SLBM deployments having foreseeably greater numbers and accuracy.

He established a contact to allow me to work with AFNIN, the Air Force intelligence organization then concerned with all aspects of strategic target numbers, selection, location and characteristics. The Analytical Services Corporation (ANSER) was also very helpful in providing target data. ANSER was far more useful to the Air

Force for urgent studies than RAND. This was a matter of their eagerness to achieve near-term results, and their location, a mile from the Pentagon versus 2,600 miles for RAND-Santa Monica. Jack Englund was the ANSER president. We usefully exchanged data and studies for nearly two decades.

I began to work on the Colonel's two problems and continued through most of 1964. One aspect of the study was to quantify the number of SRAMs needed for defense suppression and other missions by the B-52, FB-111 and later AMSA bomber forces in a general nuclear war with the USSR during 1970–1990. As mentioned earlier, over 1,000 of these missiles were later produced for the strategic bomber forces.

The strategic bomber weapon delivery missions would necessarily evolve from the then-current presumption of initial strikes on large numbers of strategic targets to BDA and selective re-strike of all targets, following the initial ballistic missile attacks. The growing numbers of ICBM and SLBM weapon delivery systems were a much more efficient and faster means of destroying fixed targets of known location. These capabilities were enhanced in later generations of missiles by much greater accuracy and the proliferation of multiple re-entry vehicles (MRVs) and multiple independently targeted re-entry vehicles (MIRVs).

Combining the factors of many more ballistic missile warheads, growing numbers of Soviet hardened targets, and the associated declining requirements for manned bomber missions, was no trivial exercise. The study provided an appreciation of the number of BDA and re-strike missions that might be required during each year of the 1970s and 1980s. These estimates helped to define the total number of bombers needed for such missions as a function of time.

The Air Force SRAM concept required very little modification of the B-52 avionics. However, ASD-Dayton was investigating the use of SRAM to increase the overall effectiveness and utility of the B-52 and SRAM through re-instrumentation of the B-52. This was to have been an internal ASD study. In deference to the pre-emptive non-funded studies by IBM and others, ASD permitted contractor participation. The IBM Re-Instrumentation Study was used by ASD engineers as a reference for study definition.

The Defense Suppression Missile (DSM) became a major funded program. Special Operational Requirement No. 212 re-designated the development as the Short Range Attack Missile (SRAM) and required investigation of the missile application to the B-52, B-58 and TFX (later the FB-111) aircraft. The utilization of SRAM for the B-58 was improbable for several technical and operational reasons. The still-extant B-70 development was never considered to be a viable carrier for SRAM.

On September 16, Arthur DuBois and I met in Dayton to see Ray Blocker, SPO Chief for the B-52. The next day we were at the AMSA/B-52 Bidder's Conference at SAC Headquarters in Omaha, NB. IBM attendees were limited since this initial meeting was intended primarily for the airframe companies. We spent some useful time with Colonel Ted Bishop who was the SAC Action Officer for these programs.

SAWSEU II

Lt. Colonel P. M. Spurrier asked me to visit to the Advanced Manned Strategic Aircraft SPO at ASD-Dayton on December 22, 1964. We met with his senior staff people, now including Jack Trenholm, the new civilian assistant SPO chief. The Air Force customarily appointed a senior civilian engineer to this role to provide continuity in long-term projects. The military SPO chief was reassigned every few years. I knew Jack from the Dynasoar Program; he had few peers in this role.

The Air Force was concerned over the slow progress in implementing the AMSA Program due to Secretary McNamara and other senior Pentagon officials. They needed a position paper from an independent source to present the logically defensible case for the AMSA development.

The Air Force believed that it would be very useful to have an integration of the original SAWSEU Paper (that had led to the selection of ERSA/AMSA in 1963) with the later work to quantify the need for B-52/SRAM and AMSA missions during 1964. They wanted the paper to educate Pentagon adversaries on the need for AMSA, or at least "to smoke them out." Later developments indicated that USAF and SAC Headquarters were also aware of this request.

We met privately for almost six hours to discuss how the new study should address the problem and to update some the technical factors and intelligence data that would be essential to the effort. They knew that this task would come on top of my existing IBM obligations, but they asked that the paper be completed in less than two months. At least the holiday season and attendant work slowdown provided time for a good start, and my new preoccupation was not apparent for two weeks. Later, FSD senior management accepted my assurances that the work would not interfere with a concurrent obligation to do an ambitious Division-wide military space study.

As in all major papers or "think pieces," the work began from First Principles. These included development of a defensible general penetration/survivability theme, a counterforce exchange scenario for both the Soviets and the United States, and a detailed definition of general nuclear war. The definition of "conclusion" of such an event drew on the Von Clausewitz definition of "prevail" — essentially, the complete disarmament of the enemy and destruction of their will to fight.

The counterforce exchange model included several thousands of the most important US and Soviet strategic targets, both current and projected. The opposing strategic offensive forces: ICBMs, SLBMs and heavy bombers, were included as they became operational. The US forces were to be launched under confirmed attack (LUCA) by the Soviets. This was necessarily a very general strategic counterforce and countervalue exchange model, unlike the detailed models that I later developed during the 1970s and 1980s.

Unlike the previous two studies, where the Air Force had allowed independent work and hoped for the best, they asked for an interim review of the draft paper. After an evening flight to Dayton on January 18, 1965, with the first draft, I spent a day at the AMSA SPO with Spurrier, Trenholm and a new staffer, Captain Emil Block. Emil's assignment was to go through the paper's quantitative aspects in great detail. He found a couple of glitches. From Emil's iconoclastic attitude and somewhat abrasive manner, it was clear that he would retire either as a major or a major general. Fortunately for the Air Force, it was the latter.

Lt. Colonel Jack Henderson had moved from the Foreign Technology Division (FTD) in Dayton to AFSC Headquarters at Andrews AFB. He agreed to review the latest paper separately. We spent five hours in discussions on January 28. As the military head of AFSC Intelligence, he had become one of the most inside of the "insiders."

As an aside, Jack mentioned that President Lyndon Johnson had gone all-out to reassure the public in his 1964 election campaign. He had declared the "missile gap" to be non-existent, on the basis of the fact that the Soviets had only 128 operational ICBMs. This number was classified above Secret at the time, because the Soviets would be very impressed by such precision and would deduce that the US had some remarkable "sources and methods." We did.

Another contribution that President Johnson made to public understanding of defense in 1963 was the revelation of the "Blackbird" aircraft, the first Mach 3 intercontinental reconnaissance aircraft. Since the reconnaissance–strike mission was now more acceptable in the professional community, the Air Force had designated it the RS-71. When he announced the program, the President mistakenly called the aircraft the SR-71, and so it has been known for the last 40 years. One does not correct the commander-in-chief. The reconnaissance capability of the SR-71 was superb, but the strike aspect of the mission was never considered seriously.

As one of the best of Air Force intelligence officers, Jack Henderson's assurance that proper use was made of all intelligence data was very important. The study purpose was always to provide new insights from existing intelligence, never to inadvertently create new intelligence. He made a few changes and verified his approval by telephone on February 2, as did Lt. Colonel Spurrier.

The final 20-page version of the report, now entitled "Mission Requirements for Strategic Aircraft," was sent to reproduction services on February 18, with plenty of disclaimers for IBM. IBM wanted all of the distance from such work that lawyers could provide. FSD Counsel Bob Moore required that any further distribution of the paper be determined by the AMSA SPO.

Progress on the FSD Military Space Study required me to work in Owego some of the time. On February 23, the new manager called from Rockville to say that he had blocked the transmittal of the paper because the matter had not been sufficiently discussed with him. I took the next plane back to Washington and spent much of the following day going over the matter, but the FSD president was out of town. This delay was no mere power play at my expense; I saw it as putting a hold on matters vital to national security. I called the AMSA SPO to apologize for the delay. My technical staff manager had an hour-long learning experience the next morning from a very irate former marine sergeant. He withdrew his objections and the report went immediately to the AMSA SPO.

On March 1, the AMSA SPO called, after their final review, to authorize the distribution of the paper to USAF and AFSC Headquarters and DOD. They sent copies to SAC. The immediate response was quite favorable, causing the SPO to release the paper for general distribution in the Commands and others such as the ANSER and RAND corporations.

Since the Air Force at all levels was now generally supportive of AMSA, constraints on the program were primarily in the upper levels of DOD. In response to an earlier initiative to DDR&E, a letter was received from Dr. Eugene G. Fubini, then with the Assistant Secretary of Defense and Deputy Director of DDR&E. He noted that the "case for strategic aircraft did not rest on assured destruction or damage limitation, but on the complete elimination of enemy residuals." He believed this to be a desirable objective but observed that "the Clausewitz definition of 'prevail' seems to be simply the reducto-ad-absurdum case for damage limitation."

Fubini's response implied that DOD senior managers were very predisposed toward much smaller bomber forces, depending mostly on ballistic missiles for a measure of assured destruction (and deterrence) and their concept of limited nuclear war. Such predispositions gave them latitude to ignore the need to finally prevail on the assumption that any nuclear war with the USSR would be of limited scope and duration. This very convenient theorizing assumed that the Soviets thought the same way about nuclear war.

The past several years of studying intelligence reports on Soviet military forces, their rapid expansion, and projected increases showed that the DOD view was either naive or convenient sophistry, allowing them to reduce expenditures on a new strategic bomber to offset the high cost of ballistic missile deployments and our increasingly expensive involvement in Viet Nam.

The Air Force had now confirmed the source of their problem. They asked for briefings to the next lower echelon in DOD to develop some converts. During the next three months, the paper was presented to many people in DOD and WESG. These included Fred Paine, Colonel Lewes, Brigadier General Glenn Kent and Carl Covington. Mr. Sturm in DDR&E Plans and Policy and Colonel Lewes (WSEG) were particularly receptive to the arguments. In all, it was a six-month effort to produce and propagate the views in "Mission Requirements for Strategic Aircraft."

Chapter 5. New Bomber Avionics

President Kennedy's plan to put a man on the moon had developed into major commitments by NASA to industry in mid-1963. The Von Braun Team at the Redstone Arsenal in Huntsville, AL, was making contractor selections based on corporate reputations and the ability to grow rapidly, make long-term commitments, and produce timely results.

In late 1963, IBM won the contract for the Instrumentation Module, a huge ring that was the last stage of the Saturn vehicles. The initial award was over $300 million and required that IBM build a major new facility for about 1,000 IBM and NASA people near the Redstone Arsenal in Huntsville.

The Owego Space Guidance Center was tasked to build, man and supervise the new Huntsville facility. This required the transfer of much technical talent from Owego to Huntsville, causing a major reduction in Owego's BNMGS capabilities.

Formation of the IBM Federal Systems Division in 1963 required development of a five-year operating plan for the division. The new FSD President, Don Spaulding, intended to ensure the accountability of the senior managers. His initial efforts were unsuccessful; the center managers preferred their prior relative autonomy. On March 3, Spaulding had me re-write the entire plan for the three business areas: Aircraft, Missiles and Space, "due to my Division-wide experience with the several business areas." This had the side effect of curtailing my consulting work for the Air Force, which IBM could readily forgo.

Each Center was provided a statement of work to produce tangible and defensible inputs for the plan. I visited each facility to ensure their understanding of the plan and the assignment of competent people. Center managements reluctantly committed to specific revenue and profit for performance measurement.

Shaping Public Opinion

In September 1964, Herman Kahn's Hudson Institute sponsored a seminar at the Hilton Inn in Tarrytown, NY. Kahn was internationally famous (or infamous) for his book, *On Nuclear War*. As one of the few attendees comfortable in opposing some of his views, the intellectually formidable Herman arranged an informal discussion with me related to general nuclear war. Professional newsmen attending the seminar were also invited.

After I contested his 41 steps of escalation to general nuclear war, he eventually admitted these were a product of "intellectual thoroughness." He also noted that *On Nuclear War* had required the collaboration of many RAND professionals. He joked that his role was putting it all together and getting the blame for it. He said that he was genuinely pained when some of his European detractors compared his book to *Mein Kampf*.

There were group discussions of the major philosophical issues of our time, including the choice of retaliation in a general nuclear war. Father Farley was a young, bright Catholic priest who was very much against wars and killing under any circumstances, even in retaliation to preserve the nation. With nearly 2,000 years of precedent to support his position, Father Farley sincerely asserted, as God's representative on Earth, that anyone opposing this position would be condemned. I noted that he was the best argument I'd ever encountered for the separation of Church and State. There was no immediate lightning bolt, but I am still waiting.

Bill Beecher, a reporter for the *Wall Street Journal*, also attended the seminar. He was the only reporter I ever considered to be both accurate and trustworthy. The Soviets and the Warsaw Pact shared this view; their reporters contacted him on special initiatives for many years.

Bill had a long and successful career as a reporter, manager and author. He later entered the government twice, in DOD and then as the official spokesman for the Nuclear Regulatory Commission. Bill received the Pulitzer Prize in Journalism for his work on US–Soviet relations and arms control.

We shared a common principled interest in broad defense issues. On September 19, 1963, I offered to assist him on an article for his paper, stressing unclassified information and non-attribution. He agreed, specifying that I would not be allowed any review of his article prior to publication.

The new information I gathered at USAF HQ, AFSC and ASD-Dayton supplemented Beecher's already considerable knowledge of these matters. Bill wrote a very comprehensive piece, bringing in the political overtones of the Johnson/Goldwater contest for the presidency and McNamara's opposition to a new bomber vis-à-vis ballistic missiles.

During 1964, Air Force consulting and planning for FSD required more than 40 trips to the major Air Force commands. At Dulles Airport, one time, I saw Jimmy Hoffa sitting, seemingly, alone. Walking toward him, I was suddenly confronted by a very large, menacing individual. Hoffa motioned his gorilla aside, and we talked for about 20 minutes regarding unions, Communism, and how to run an organization with two million members. I mentioned that back when I was a General Electric radar technician, there had been a serious run-in with a Communist-influenced unionization effort. He recalled a similar experience with Communists in a Pontiac, MI, factory, but said he "just sent a bunch of guys up there to break their heads."

On January 28, 1965, Don Spaulding was replaced as FSD President by Bob Evans. Bob was bright, experienced, vigorous, and had developed a very decisive management style. He asked to be briefed on current military intelligence, the B-52, SRAM, AMSA and the Military Space Study. He had been well informed regarding my work in Owego and the volunteer work for the Air Force.

The Air Force was finally preparing to issue a request for proposals to define the AMSA Avionics Development Program. Bob Evans and John Jackson had decided to give me a temporary assignment in Owego to manage the proposal effort, and if we were successful in winning the study, a permanent transfer there. Evans gave me the afternoon to consider it, but he wanted an affirmative answer by 4:30 p.m.

AMSA Avionics Proposal

Arthur DuBois and I reviewed our tentative plans with Evans on May 18. On May 27, John Jackson asked for assistance in the AMSA Proposal. After the AMSA Avionics Subsystem for Strategic Bombers (ASSB) Program Request for Proposal (RFP) was delivered, we worked 100 hours during each of the next four weeks.

The ASSB Proposal Team consisted of 15 to 20 professionals from operations research, systems and sensor engineering. We proposed an AMSA bombing, navigation and weapon delivery system very similar to the configuration designed for the B-52 Re-Instrumentation Study three years earlier. All of the subsystems were upgraded to employ the newest equipment that had been developed in the interim years.

We emphasized the need for thorough study of the Soviet Air Defense System (ADS) and their massive, hardened strategic target structure that would have to be destroyed to allow the US to prevail in a general nuclear war. This emphasis on threat synthesis and mission profiles justified major study expansions during later contract negotiations.

The management proposal stated that the study manager would report to the Owego manager of engineering. This was done to remind the SPO of the previous three years of voluntary work on their behalf and to discourage Jackson's deciding otherwise after a possible contract award to FSD.

Paul Hockman, a very experienced avionics system manager with a jaundiced view of contractors, was the senior civilian manager for AMSA Avionics. The proposals submitted by NAA-Autonetics and IBM were superior, so only these two companies were selected for the competitive studies. In later years, Paul said that the IBM submittal was the best avionics system proposal he had ever received. The ASSB Contract was the first major avionics system study awarded to FSD-Owego in seven years.

After the unofficial Air Force decision for contract awards to NAA-Autonetics and IBM-FSD was known to the industry, Bob Evans ensured my promotion to functional manager (3rd level, line manager) and placement on the IBM corporate payroll, after 11 years in staff positions through Senior Engineer. The transfer to Owego was effective in August 1965.

As manager of strategic aircraft programs, my job description was essentially that proposed to Art Cooper and Clint Grace three years earlier. Del Babb was then manager of engineering in Owego. Del was a big, pleasant man, but reluctant to build the organization needed to perform the new ASSB Study. Our initial cadre was ten professionals. Still, with Art DuBois' invaluable assistance and current knowledge of the best people in the facility, we built an organization capable of timely performance on the ASSB Program.

IBM was selected to work with Air Force intelligence offices to identify the projected Soviet ADS threats for both the IBM and NAA study groups. An adjunct study was originated for AMSA active lethal defense, in addition to the infrared and electromagnetic sensors and transmitters needed for passive defenses. These additions increased our initial contract scope to $950,000 for the ASSB Phase I Study. Six phases and many separate experiments were negotiated during 1965–1970, for a total of $12 million.

We needed a manager for controls and displays, so we asked for John Cooney. Jackson finally told him to transfer. John proved to be worth all of our efforts and eventually forgave our insistence.

By normal industry standards, we were always overworked and understaffed. The dedication, abilities and determination of the professionals in our organization made up the difference in achieving superior performance.

NAA-Autonetics had three times as many people on their study than IBM. They recognized the ASSB Study as a vital, enduring competitive effort and provided corporate funding far beyond their contractual funds. Their reports were twice as lengthy as ours. However, the SPO people advised us that quality and conciseness made our work preferable to Autonetic's.

The intense 10-month effort provided major reports during April–June 1966. Volume I described the characteristics of all specific ADS weapon systems that posed a threat to AMSA during deep penetrations into the USSR. These included current and projected anti-aircraft artillery, surface-to-air missiles (SAMs), and interceptor aircraft armed with guns and air-to-air missiles (AAMs).

Volume II deployed the foregoing ADS weapon systems to develop a high-altitude threat environment. Volume III described the same deployments for the low altitude threat. These documents were submitted in June and re-issued in October, 1966, along with comparable threat environmental models for limited warfare in Central Europe and Southeast Asia.

The foregoing early definitions of the anticipated ADS threat environment were necessary to establish design requirements for the phased-array forward and aft radars, infrared sensors, and the air-to-air missiles (AAMs) needed for lethal defense of the AMSA. The aircraft was designed to prevail in its mission as a single weapon system. These threat projections were also necessary for design of the electronic countermeasures (ECM) sensors and transmitters for passive defense.

The early Phase I studies addressed several essential tasks: (1) formulation of required ASSB capabilities; (2) equipment performance criteria, sensor studies and trade-off analyses; (3) quantitative error analysis, and (5) identification of advanced development tasks. In the latter category we identified what became Experiment 5, related to ECM, and Experiment 6, Integrated Controls and Displays, both of which were awarded to IBM in later competitions.

Bob Evans decided that Jackson should have an assistant general manager. L. Michael Weeks was hired for this position from MacDonald-Douglas on November 28. He immediately required that we spend several hours each day in meetings in his office. This meant working even longer hours to fulfill the obligations of Department 566.

Mike injected himself into all Department 566 responsibilities. As an FSD Senior Manager, this was his right, but he lacked the background, ability and judgment for this role. Owego Marketing encouraged this involvement.

ASSB Program Growth

ASSB contractual obligations allowed our organization to grow to 30 people by May 1966, with peripheral support from other engineering functions. Owego management finally recognized that the future of the IBM avionics business would be determined by our functional organization.

Our best near-term new business growth would be the ASSB experiments identified by NAA and IBM. We concentrated on the Soviet Air Defense threat studies and lethal defense. Later on, we stressed computer requirements, controls and displays, avionics system simulation and electronic countermeasures.

All of this proposal and contractual activity was interspersed with many meetings with local senior management, Bob Evans, Thomas J. Watson, the AMSA SPO, Air Force Oklahoma City Air Materiel Command, SAC, USAF Headquarters, and dozens of avionics contractors. We generated over 100 major reports and proposal documents.

Experiment Five

FSD-Owego had a formative ECM capability in Ed Zaucha's large functional organization, but they had not yet won a significant development program. Experiment Five of the ASSB Program was intended to define ECM system requirements, responsive technology, an ECM system configuration, and the hardware "breadboard" to demonstrate critical functions for the AMSA ECM Subsystem. During early May 1966, we prepared the technical, management and cost proposals needed to win this contract.

The ASSB threat studies gave our team an appreciation of the great numbers and diversity of the signals radiated by the Soviet Air Defense System. These threats would have to be received, sorted and prioritized in real time at the various altitudes and supersonic speeds of the AMSA.

The ECM transmitters needed to be tuned to selectively and immediately jam these signals over broad frequency bands — an equally demanding task. Very sophisticated and expensive "comb filters" made from yittrium garnate (YIG) just being developed by other companies would to help perform the signal identification and sorting task. Zaucha's organization was anxious to win and the final price was not their responsibility. They provided low estimates for the YIG filters; this later caused many cost overruns and organizational problems.

Since our ASSB organization wrote the proposal as an integral part of our overall program responsibilities, we were listed as the managers of Experiment Five. Jackson, without our knowledge, changed the program responsibility to Zaucha, and Marketing took the new version of the proposal to the AMSA SPO.

We were awarded ASSB Experiment Five Contract for $2,300,000 on July 25. The following day I received a congratulatory telegram from Bob Evans.

With Zaucha as the new program manager, I could not insist that the work be performed as scheduled or withdraw the funds. His organization randomly worked on Experiment Five for a month. But since the ECM system would be vital to the survival of the AMSA Aircraft (and thus the entire AMSA Program), the work had to progress on the proposed schedule. After delays and many complaints, I advised Jackson that if he was not going to perform on the program, perhaps he should "give it back to the Air Force for re-competition." John was livid, but he got the point.

Zaucha's large organization had discretionary funding and a capital equipment budget. He committed these funds as deemed desirable and legally applicable. This allowed presenting a credible position to the AMSA SPO during the technical negotiations on August 29.

We did note that our bid was optimistic with regard to the costs and had provided extra funds — and asked if the SPO could be equally forthcoming. They reduced the work scope and allocated additional funding. We re-wrote the IBM work statement at 3:30 a.m. on August 30 in order to conclude the negotiations.

Zaucha's ECM group continued on for six months without producing any tangible work product. We could no longer placate the AMSA SPO through superior technical performance on the main ASSB Program. We forced a new DCA meeting and the participants refused to give Zaucha any more supplemental funding for Task Five. The AMSA SPO was so advised on March 30.

The internal contention culminated in a very unpleasant meeting with Weeks and Jackson from 11:00 p.m. to 2:00 a.m. on April 10. We chartered a plane at 4:30 a.m. to spend the day discussing this mess with Paul Hockman and the SPO Chief. They were more understanding than we had a right to expect, in part due to the many services to the AMSA Program during 1963–1965 and our superior performance for the past two years on the main ASSB Study.

Lt. Col Ludlow called on April 12 to say that our present contract was cancelled and that the AMSA SPO would rerun the Task Five competition. They would let IBM bid again, rather than exclude us due to non-performance.

The second Task Five bidders' conference was held at ASD-Dayton on April 18. John Jackson attended; he wanted to win the program again to mitigate the stigma of non-performance, for which he was primarily responsible.

Our second Task Five Proposal was sufficiently competitive to warrant technical negotiations on May 9. On May 24, Jackson and Weeks informed us that they were going to give Task Five to Zaucha again if the program was won a second time. I found this arrangement unacceptable and asked for a group visit with the FSD President, Bob Evans, to reconsider this arrangement. Four hours with Evans and Weeks were of no avail.

We re-won Task Five on June 15, with the contract value increased to $2,700,000. This time Jackson gave Zaucha the whole package, ensuring little chance of a successful program — and this was integral to our ASSB effort.

Experiment Six: Simulation Laboratory

The AMSA (B-1) avionics was so complex and diverse that a full configuration would weigh about five tons. Control of the avionics would require two operators seated side-by-side in front of two large arrays of controls and displays. One operator would control navigation, bombing and missile launching functions. The other operator would primarily control the ECM and lethal defense. Due to varying workloads for different phases of the mission profile, inter-operability between the two crewmen was also required. Essential displays were separately provided for the pilot and co-pilot.

A Statement of Work (SOW) for Experiment Six was received from the AMSA SPO on July 1. We prepared a plan for a major simulator development; discussed Experiment Six ground rules and scope with R.E. Porter in Dayton and met the new SPO Chief, Lt. Colonel Ludlow; reworked our Experiment Six Technical Proposal and submitted it to ASD-Dayton; and negotiated the contract on September 2. The final technical and cost proposals were submitted on October 10. The First Interim Technical Report on AMSA Avionics Development Task Six was submitted on December 15, 1966.

The simulation facility had been very important to FSD in the early years of the B-52 and B-70 to develop controls and displays for their complex avionics systems. By this time, the laboratory had deteriorated to a directionless hobby shop for Human

Factors and other engineering functions. The laboratory was transferred to Department 566 and assigned to John Cooney due to his controls and displays responsibilities on the ASSB Study. He used the $2,000,000 capital equipment budget proposed on Experiment Six, now called Task Six, to completely rebuild and refurbish the laboratory.

This installation provided a two-station simulator for the AMSA Navigation–Weapon Delivery and Defensive Subsystems Operators. John's department grew to over 100 people and consequently Department 566 approached a total of 200 engineers, physicists, mathematicians, operations analysts, psychologists, programmers and program management/control personnel.

John Cooney rapidly transformed the laboratory into a vital engineering tool for AMSA and A-7 Proposal studies. His organization produced four high-quality Interim reports on ASSB Controls and Displays during January–October 1967.

We had managed to grow Department 566 by 2,000 percent in 18 months.

Tactical Avionics Programs

FSD-Owego had failed to capitalize on any of the market opportunities developed for them during 1960–1963. In 1964, a diversification into avionics subsystems for tactical aircraft could still be a solution to Owego's declining posture in military avionics. The best early program prospect was the TFX, a new tactical fighter-bomber later called the F-111. Owego submitted an unsuccessful proposal. Hughes Aircraft (Avionics) and others won the study and development contracts.

Owego Computer Engineering was developing, with the assistance of IBM Commercial technology (and the usual stimulation from Bob Evans), a new line of airborne digital computers called the 4-Pi. Modular design of the computers could be adapted to a spectrum of possible applications. By using components from the massive production runs for the IBM 360 Series of commercial computers, such as ferrite core memories and layered circuit boards, FSD could eventually underbid all of the competition for 4-Pi applications. As yet, they had not succeeded.

Having failed to win one of the F-111 Avionics Integration contracts, FSD-Owego used their Computer Development Function to prepare and submit a computer-only development and production proposal for the F-111 to Hughes Aircraft (HAC). This singular bid for the F-111 Computer was submitted in late 1965.

On December 8, 1965, a marketing representative, Earl Sparks, and Jim Crenshaw asked to meet with me. They were very concerned that FSD should also propose the F-111 Computer directly to General Dynamics-Ft. Worth. This would eliminate the loading imposed by HAC as the avionics integrator and demonstrate to General Dynamics-Ft. Worth the cost advantages of a direct purchase from IBM-Owego. John Jackson had told the Mark II proposal manager(s) to prepare such an alternate proposal. However, they ignored his direction. I agreed with Sparks and Crenshaw, so I met with his current proposal management and told them that if they would not submit a direct proposal to GD-Ft. Worth, Department 566 would, together with Computer Development. Babb agreed and Jim Crenshaw transferred to our department on December 16.

On December 23, the new team began proposal preparation. We now had overall responsibility for the effort, with Jim Crenshaw as the Mark II Proposal Manager.

When GD-Ft. Worth decided to eliminate the potential F-111 avionics integration contractors and perform the entire avionics integration task themselves, causing the original bid to HAC to be useless. Now our direct bid to GD-Ft. Worth was the only remaining near-term hope for tactical avionics computer applications.

There were some very long days and nights with Crenshaw during March 12–27. A total market potential of 1500 Mark II computers for the F-111 with spares and support caused our final bid to be about $90,000,000.

Bob Evans was the first FSD president to become personally involved in competitive contract negotiations. He recognized the importance of this bid, and insisted on severe cost reductions to be attributed to the advantage derived from using many IBM commercial components; this ensured that the Mark II Computer Contract would be awarded to IBM.

When we assumed responsibility for the alternative Mark II Computer Proposal, Department 566 became Advanced Tactical and Strategic Aircraft Programs. That was all the recognition we received for the Mark II contract award. I eventually managed to secure an award for Jim Crenshaw as the Mark II Proposal Manager, but Owego senior management and marketing liked to think that it was they who won new programs. Credit for failures was reserved for functional managers.

Chapter 6. A-7D/E Avionics Program

By winning and rapidly expanding the AMSA Avionics study (ASSB) and winning the Mark II Computer, my team in Department 566 had thoroughly proven its entrepreneurial effectiveness. Owego senior management began to shift this responsibility to our functional organization. But the business prospects of Owego remained bleak overall. Arthur DuBois and I were very concerned about FSD-Owego's rapidly deteriorating posture as a major avionics integration and production contractor.

Winning the Mark II Computer production program by mid-1966 alleviated some of the concerns and assured the future of that business area. The necessity to win a new major avionics system integration contract while performing on the ASSB Program, and cleaning up the mess Owego's management had made of ASSB Task Five (ECM), were continuing preoccupations. Since the contract for AMSA Avionics development and was two or more years in the future, we might not be competitive for the final ASSB contract.

The A-7D/E Avionics Program was the biggest (and only) timely tactical avionics contract award for 1966–1967. The Sperry-Rand Corporation had three years and several million dollars of US Navy funding to study and design the new avionics system. It would be a very difficult for Department 566 to win.

Owego senior management now had a win–win situation by giving this responsibility to Department 566. If we won, great; if not, they could remove the department manager and still have the ASSB Study Program and Mark II production contract — a $100,000,000 improvement over mid-1965.

A-7D/E Avionics Proposals

The added responsibilities of Department 566 were crushing. Department L04 was not transferred to us until August 6. L04 was managed by Bob Wall and included Jim Kiernan and Lenny Masowski. They were all very good, dedicated engineers but had lacked upper management direction and support for many months.

We formulated a preliminary bidding posture for an A-7D/E Avionics Proposal to the Ling-Temco-Vought (LTV) Company in Irving, TX. There was an exploratory visit to LTV-Irving on October 20. Since IBM was late in joining the competition, we would need Pentagon and Navy approval to participate.

We began immediate preparation of a technical proposal for a DCA meeting on November 23. On Thanksgiving Day I had the first full night's sleep in three months.

As the LTV Director of Procurement and Contracts for the A-7D/E, Bob Henry was the quintessential Texas good ole boy who knew what he wanted in an avionics system contractor. We developed a very good personal and working relationship over the ensuing 11 months.

Sperry-Rand considered themselves to be unbeatable and behaved accordingly, doing end-runs on LTV, thus offending Bob Henry. LTV engineers knew relatively little about the new digital avionics computers. Sperry-Rand was exploiting this circumstance to ensure a larger role in the A-7D/E Program. We suggested to Bob that IBM conduct a series of educational seminars to correct this problem. He agreed.

Bill Carrol was the lead engineer for the A-7D/E Computer in FSD. He was a mature, conservative computer engineer, assisted by Fred Jenny. Fred was a very creative designer inclined toward higher technical risks. They made a good, balanced and competitive computer design team.

Carrol, as a project manager in computer development, was the obvious and willing choice to be the FSD A-7D/E program manager. He was always described in that role in our proposal documents. Should we win the competition, we expected all of the proposal engineers in Department 566 to be transferred to this new program, as had been the case with Crenshaw on the Mark II computer contract.

A proposal manager needs a much different mind-set and greater energy level than a program manager. Critical decisions must often be made within hours on a proposal; a program manager may have months to respond.

Owego Marketing had always been an ineffectual function — that was why Owego was going broke. I was sent there to fix it, and made it clear passively and occasionally very specifically that my group would perform many of the normal marketing functions, especially business strategy and important customer relationships with both military and industrial organizations. Joe Jennette, Jr., was Owego's manager of marketing. Ken Driessen now resided in Owego and was assigned as the marketeer responsible for our business area.

Having overcome Marketing's failures and Owego senior management's lack of leadership on the ASSB and Mark II programs, this was again necessary for the A-7D/E Program. The simple choice was whether to please them or feed them and I again chose the latter. Had we ever failed to win, they would have promptly found a new entrepreneurial functional manager.

Competition

There were 13 firms participating in the A-7 Avionics competition; every US company in the avionics system industry wanted to win it.

Our first four-volume proposal was submitted on November 28, 1966. Initially, our avionics computer provided only a 4,000-word memory and an operational speed of 50,000 operations per second (50 KOPS).

ESC-Owego, renamed the Electronics System Center, had 15 years of experience in high and low, straight and level, strategic bombing computations. We knew very little about the tactical, high-speed, maneuvering weapon delivery of cannon fire, rockets and bombs. In early January, we asked for $50,000 (two man-years) to develop a set of tactical bombing equations. The money was forthcoming but not the mathematicians. Two months later, in matters related to AMSA, we absorbed the Owego mathematicians including Drs. Vanderkulk, Standish and Budurka. Then Jackson took the money away. Apparently, senior management thought that our chances of beating Sperry-Rand were minimal. Further, this would ensure a loss for which the proposal manager could be blamed and fired.

LTV reduced the field to seven companies for the next phase of the competition. In two weeks our four volume proposal described a central computer with an 8,000-word memory and capable of 75 KOPS. Some of the proposal funds and then the mathematicians had helped to improve our early bombing equations.

Our A-7 proposal efforts caused an unprecedented visit by both the Navy and Air Force on March 7. They asked for an Associate Contractor Proposal, separate from our LTV efforts. Another four-volume set of proposals, to be produced concurrently with our primary competitive effort, described how and at what cost we could provide the A-7D/E Avionics independent of LTV. We briefed our proposals at the Naval Aeronautical Systems Center on May 5.

Senior management was now less pessimistic. Funds became available to send all of our PhD mathematicians to any facility that could be helpful, particularly the test facility at China Lake, CA. They were very effective in eliminating this major fault in our competitive posture — just in time for a Navy/LTV visit to ESC on May 10.

Final Phase

LTV finally issued a system RFP on April 24, requiring System Support and Supplemental proposals on May 1 and May 22 respectively. Captain Doss (USN) ordered a final competition for only three contractors.

Fully appreciating the weapon delivery and avionics subsystems management problems warranted a final computer configuration that included 16,000 words of memory and an operational speed of 125 KOPS.

On May 23, Jackson and Weeks asserted again that Zaucha would manage Task Five and that Carroll and Crenshaw would run the A-7 Avionics Program. They also edicted the eleven principal managers selected to run the anticipated A-7 Program. The hard work was nearing completion on the A-7 proposals; win or lose, the long knives were coming out.

The A-7 Avionics contract was expected in early June, but the situation remained fluid until mid-August. Needing experienced proposal leadership and my good working relationship with Bob Henry in LTV Contracts, Jackson and Weeks delayed taking their preferred initiative.

I called Evans from LTV-Dallas on June 3 to review the lessons learned in our May 26 meeting. Mike Weeks' involvement would jeopardize our chances of winning. Jackson was a much more able manager, so I asked Evans to intervene. From this point onward, I only talked with Jackson and Driessen when determining our necessary actions during the competition. Evans complained about my strained interpersonal relationships at Owego. He had also assumed the ASSB Program was my only responsibility. He was allegedly unaware of who was behind the successful Mark II bid or who was now also responsible for the A-7D/E competition.

Our final, formal proposal documents were submitted to LTV in mid-June. Many interim machinations regarding the A-7 Avionics procurement by the Navy, Air Force, DOD/DDR&E, Sperry-Rand and LTV were avoided by simply supplying data and support to LTV.

DDR&E had negated the first attempted award to IBM, although the choice of LTV had been based on strict evaluation of the technical, management and cost proposals. DDR&E favored Sperry-Rand due to their three-year sponsorship of the ILAAS Program. They could not accept the fact that IBM had defined and proposed a better avionics program in just ten months.

After working with our FSD Proposal Team for a few months, LTV had decided that, at all management levels, IBM would provide a better A-7 Avionics Program and be easier to work with than Sperry-Rand. LTV senior and contracts management had also decided that they would prefer to have the proposal manager as the IBM A-7D/E program manager for at least the first 18 months.

Bob Henry privately advised me that LTV did not ordinarily specify a particular manager to their subcontractors; this would be the first time in many years. This was very troublesome, since I didn't want the job — and Owego senior management would be apoplectic.

Negotiations

In mid-July, Bob Henry asked that all of the principal members of our A-7 Proposal Team come to Dallas for several days of discussion and negotiations. We were nearing a resolution of the A-7 competition and this would be our last chance to win (again). LTV had formulated the A-7D/E Avionics Program that they believed would overcome any objections from DDR&E.

We had three days of very constructive discussions, with specialists from both IBM and LTV pairing off to try to accommodate the new program criteria. By July 14, almost all issues were on the way to being satisfactorily resolved.

Bob Henry and I decided on our next step. Since we now agreed that IBM could meet LTV's terms, we set up a final meeting, with the lone LTV negotiator, at 3:00 a.m. Sunday. This would imply that he had taken on all 13 members of the IBM negotiating team in the wee hours and had prevailed on LTV's terms and costs. He agreed that it would be "good theater." After some discussion with Henry, and document exchanges, IBM "capitulated" to successfully complete the negotiations. We pleasantly parted company at sunrise.

The following week, Bob called on yet another major item. He needed the number and types of people and the attendant costs for in-plant IBM offices at our subcontractor locations — by midnight.

The relevant members of the proposal team were gathered and we added Jim Crenshaw and Del Babb to benefit from their experiences with the B-52 and B-70 procurements. In four hours, we identified 30 people and several million dollars in associated three-year costs. We put in a conference call to Bob Henry before midnight and unilaterally committed IBM to do this task at a firm, fixed price. All this was without senior management involvement and the required DCA proceedings — easily a job-loser if the estimates were wrong. This episode exemplified the differences between proposal and program management.

On Saturday, August 5, Bob Henry called to advise that LTV would go forward to the A-7 SPO on Monday, recommending IBM as their integration subcontractor (ISC) for the A-7 Program. He called again three days later to say that the several government offices had accepted IBM as their ISC. He said that we had won the competition again, and had a high confidence in the contract award.

I advised Jackson and Bob Quinn, then ESC controller, of the A-7 Program award and that I would be making an immediate departure for the AMSA SPO in Dayton.

Driessen was also advised of the event so he could make transportation arrangements. LTV Procurement later called Jackson to confirm the intended award to IBM.

When I arrived in Pittsburgh two hours later, the airline personnel advised me I had been booked on the same plane back to Binghamton in 20 minutes. Jackson asserted that I had to return and help with the preparations for the LTV negotiations.

Calling Bob Henry from home, as prearranged, at 11:30 p.m., I learned that a new problem had arisen. DDR&E had just asked LTV to justify the A-7 Avionics award to IBM. One such last-minute briefing, just a month earlier on July 3, had resulted in our loss of the program.

I met with Jackson at 8:00 a.m. and told him the news. Stoughton and Carroll called during our meeting. Jackson said they had intended the meeting to congratulate the principal contributors on the A-7 Award. Actually, the occasion had been planned to announce cutting Department 566 by 80 percent and redistributing the assets to them. Jackson could hardly wait to effect this demotion, but Henry's new input deferred his initiative for a couple of weeks.

Henry soon called, asking for further immediate final negotiations. On August 10, we split into our usual discussion groups with LTV counterparts. At 11:00 a.m., Bob Henry came in and congratulated us on winning the program, and pointedly shook hands with me instead of Bill Carroll to make the announcement. Bob then insisted on a private meeting. He recounted the preference for the proposal manager, at all levels of LTV management, as their IBM A-7 avionics program manager. I gave him no encouragement.

We arranged a celebration dinner for 18 LTV and IBM participants including Bob Henry, George Love and others in LTV Contracts. Henry was assured that we would be a successful ISC due to the high level of interest by IBM senior management and the considerable resources at their disposal. Further, Carroll and Crenshaw constituted a very good management team.

I called John Jackson the next morning to ensure his clear understanding of my discussion with Bob Henry. Jackson did not attempt to understand. He asserted three times that I had acted unilaterally. Further, I had not communicated with others in the group, especially Mr. Carroll.

That disturbing conversation warranted a call to the FSD president to advise him that Department 566 would likely be broken up after the A-7 Award and that I had not the slightest interest in managing the A-7 Program, as LTV would prefer.

Evans asked for a summary of the mission and accomplishments of Department 566. The IBM Proposal Team developed a 30-item plan for the LTV/IBM initial program activity, the mutual steps necessary to implement the A-7 Avionics Program. He also advised me that Jackson had talked with Sol Love. The next day, Dean Thomas of LTV engineering privately related some of what had transpired.

Jackson had arrived in the late afternoon and had been met at the airport by Driessen and perhaps others. They went directly to LTV and talked with H. E. Wellborne, director of LTV procurement, and Sol Love. About 7:00 p.m., Jackson came to the room and said that he had prevailed in securing the A-7 organization that he preferred and that all Department L04 personnel must be transferred from Department 566 to Crenshaw, who was now systems engineering manager.

We met Saturday morning to effect the transition from our A-7 proposals to the A-7 Avionics Program. I thanked the LTV technical people for their part in selecting IBM and introduced our new program and engineering management. I wished them all good luck in their new joint venture.

Bob Henry walked me to my car and expressed concern about my future with IBM now that the A-7 Program had been won — and given to others. However, in the euphoria and great relief at finally succeeding in the A-7 competition and being rid of the whole thing at last, on whatever terms, I did not yet fully appreciate what he and Thomas were saying.

On August 14, the memorandum regarding Department 566 that I had promised to Evans was written: "ESC Advanced Strategic and Tactical Aircraft Program Function." I covered our charter and its inherent conflict with other functions, my focus on being selective in new business endeavors, and how I had minimized entrepreneurial expenses by covering personnel costs under the AMSA contracts. Putting the department back to work on contacts with billable hours, directly charged to the Air Force, was key to salvaging the department's finances.

The difficulties of Department 566 over the past 22 months were also reviewed. Having won the ASSB Study, Mark II Computer and the A-7D/E Program, for a total of $300 million, there would now be a respite in which to rectify some of the attendant difficulties.

John Jackson and I went to IBM World Headquarters at Armonk, NY, on August 15. We had two meetings with Dr. Eugene Fubini, IBM's new chief scientist. During the first session, we answered all of his questions regarding "buying in" and later recovering costs through engineering change procedures.

During our afternoon meeting, Dr. Fubini called DDR&E and, with passive assistance from John, finally assured them that the IBM financial position was valid as proposed and that we were able to perform as scheduled. This call finally put DDR&E concerns to rest.

When I returned from a short vacation in the Adirondacks on August 23, I placed a purely social call from home to Bob Henry. Consistent with our rapport of the last 11 months, he wanted advice on how to handle a problem with the selected inertial platform vendor. We evolved a plan that would not interfere with the normal interface between LTV, IBM and the vendor. I was obliged to relate this conversation with Henry to John Jackson; there was no reaction from John.

Jim Crenshaw subsequently asked if anyone had been transferred from Department L04. I said "No," but didn't go into details. Preparations were underway for merit salary increases, personnel appraisals and promotions, all of which were either required by IBM personnel policies or in the best interests of the individuals. Jackson and Weeks had been informed by letter on August 18 regarding this activity. I had also spent an hour with Weeks to obtain his approval of the recommended salary increases.

On Notice

I met with Mike Weeks and Jackson on August 24. Weeks stated their appreciation of my contributions to the A-7 Program and actions ensuring that the contributors received prompt recognition. He then asserted charges of insubordination based on my recent call to Bob Henry and on supposedly trying to secure a position in A-7 Program management while jeopardizing the interests of IBM and making derogatory statements about the Company. He also stated that I had failed to be responsive in transferring people to the A-7 Program in spite of a direct edict by Jackson in Dallas. They angrily refused to put their charges in writing.

They wanted me out of IBM. Further, everyone but the very minimum of the AMSA ASSB Program staff would be removed from Department 566 and would report to W. T. Chow (a West Coast acquaintance of Mike Weeks hired a few months previously).

This groundless and deliberate misinterpretation of recent events was too ridiculous and unexpected for an immediate response. As the functional manager responsible for the successful proposals and liaison necessary to finally win a program worth well over $150 million, I had thought the contract award was a considerable achievement. The A-7D/E was the first major avionics integration award to ESC in more than ten years. We had won over seemingly insurmountable odds and for the first time ever in open competition against all of the major companies in the avionics industry.

Jackson and Weeks recommended Bob Evans dismiss me from IBM. Instead he put me on "on notice" status for the maximum of 120 days — the bad cop, good cop treatment. "On notice" status was a tenuous refuge for unsatisfactory IBM employees. If the individual did anything inappropriate during the stated interval of 30 to 120 days, he could be immediately fired without recourse.

Jackson and Weeks kept up the pressure to have me fired.

Evans had written a three-page letter describing many alleged offenses to the Company, including all of my associates, specified the intended assignment of Department 566 to Wen Chow, my attendant demotion, and the statement of being on notice for the next 120 days.

Since these misfortunes would be rapidly communicated through ESC, I advised personnel in Department 566 that there would be severe organizational changes. They were cautioned against any partisan actions or discussion of the matter.

The following day, Bob Evans talked for an hour about my alternatives. I could be dismissed, voluntarily leave IBM, accept transfer to a staff position in Gaithersburg's new Space Systems Center, stay at ESC under the conditions stated in his letter, or ask his assistance in finding a position in another company. He required an answer the next day.

I called Jack Trenholm at the AMSA SPO to discuss the new organization and determine if this change was acceptable, without giving any rationale. We had mutual confidence that John Cooney and Al Johnson could still independently discharge our current and anticipated obligations.

Trenholm believed that IBM would provide better study results and prepare a more competitive posture with the current ASSB program manager. I assured him that if subsequent performance was inadequate, FSD senior management would be so advised.

On August 31, I told Evans I had decided to remain at ESC-Owego under "on notice" status — for private reasons, and gave him a better understanding of my intentions and professional circumstances during the past two years, particularly 1967. This did not mitigate the conditions of his letter, but served my purpose of staying in Owego.

Evans had essentially given Department 566 to Wen Chow. This meant I would be reporting to him following my demotion. Evans' letter specified that I was not to contact LTV under any circumstance. I was required to take a three-week "working vacation" immediately, followed by a visit to the ESC-Owego physician, Dr. Dietrich. My only remaining responsibility was to restore a competitive posture to the AMSA Avionics Program. Since two thirds of the money and people on the program were now dispersed to others, this would difficult.

Open Door Policy

The IBM Managers Manual on Employee Relations, Index 8-01, of March 30, 1967, noted that "[The appeals] policy is deeply ingrained in IBM's history. This policy is a reflection of our belief in the dignity of the individual and his right to appeal the actions or those for whom he works." The policy allowed appeals by the employee to all levels of senior management, including the IBM chairman — at that time, Thomas J. Watson, Jr.

The fact that John Jackson and Bob Evans took their unwarranted and aggressive action in spite of their knowledge and experience with this policy convinced me that such recourse was most likely not available to a now-demoted functional manager. IBM had many unwritten rules regarding such matters. Ultimately, a large industrial organization is a totalitarian autocracy in which the current situation would be of little moment. Use of the Open Door Policy likely would have resulted in my being fired from IBM immediately.

Having made the critical difference in bringing $300 million in new business into ESC-Owego in the past 20 months, I was now disposable. My only value to them now was the rapport I had with the Air Force due to work on the B-52 and AMSA. The latter program had an avionics business potential considerably over a billion dollars.

One of my reasons for accepting the demotion was to enable my wife Barbara and me to have our first real vacation in several years. Evans worked out an arrangement for me to do insignificant errands in Los Angeles and Honolulu, likely to allow favorable accounting procedures for the enforced "vacation."

The next requirement was the obligatory visit to Dr. Dietrich. When faced with an employee who did not embrace their more enlightened views, senior managers in industry would sometimes require the person to seek psychiatric help, with the then-attendant stigma. Dr. Dietrich and I had a common understanding that went back to my earlier tour in Owego under Art Cooper and Clint Grace. We both understood that somewhat manic work tendencies were necessary for survival and accomplishment in the Owego professional environment. I never imposed this attitude on other IBM employees, except by proximity and example. We usually enjoyed a cup of coffee together and sometimes a humorous recounting of the reasons for being there. This time, there was no levity.

Survival

During my pause to reflect in Lihue, Hawaii, I formulated both longer range and interim plans. Two more years of IBM service were required to acquire permanent vested rights for IBM retirement benefits; leaving now would provide nothing but nominal severance pay.

In the interim, my intent was to rectify the injustices, and failing in this, to "open door" with Tom Watson after November 11, 1969, against all of the IBM managers involved. This would likely fail, necessitating a civil legal action, however futile that might be against "Big Blue." The prospects for this alternative were so dismal that a better solution had to be developed.

I told Wen Chow that prior criticism of my communication skills and my precarious professional circumstances required that we maintain a daily or weekly summary of our every significant action or conversation for the foreseeable future. This practice would make Weeks and especially Jackson a bit wary.

The Chow file runs from September 1, 1967 through April 18, 1968, long after the on notice" status expired.

In the interim I followed Evans' AMSA-only edict. On several occasions, I refused any other activity. Such initiatives would require frequent interface with Jackson and Weeks, which I avoided.

Since edicted transfers reduced Department 566 from 200 people who had been reporting to me, directly or indirectly, to 30, I had plenty of time for my next private initiative: a detailed 40-page reconstruction of all known management actions regarding LTV-Dallas, Owego and FSD Headquarters that led to my "on notice" status. This would help to me understand what had really happened and who was most likely responsible; provide the basis for an Open Door initiative in late 1969, if necessary; and prepare documentation that Barbara could provide to IBM Corporate Headquarters in case I died or became disabled in the interim.

My continued presence in Owego was unwelcome to Jackson and Weeks. I made them still more uncomfortable during the three required "on notice" meetings, without provoking a severe reaction. Weeks' limitations made him the better target, so I tested potentially provocative thoughts on him first.

On October 25, in the required "on notice" interview, I summarized my interim accomplishments and indicated that the exceptionally excessive overtime June 1965 through August 1967 was not a result of my personal predisposition but of the excessively demanding professional and business environment.

Late one evening after our return from Hawaii, Bob Henry called. I told him that if he was calling about the A-7 Program, we could not talk. Further, I would have to inform Owego's senior management that he had called. He said that he was fully aware of the situation and this was strictly a personal matter. Henry said that he and LTV senior management all knew what Jackson was going to do after the contract was awarded to IBM. And he said, "So we saved you a job. Do you want it?" I thanked him for LTV's consideration, but advised him I intended to stay with IBM and find a way to get paid for what I had already done for Owego during the past two years.

While I was in Hawaii, Mike Weeks, in his role as assistant general manager, had visited the AMSA SPO. He told Harold Hussey, our senior marketeer in Dayton, that I was "on notice." The AMSA SPO had observed the formality, but when Weeks presumed to be their new interface with IBM, Paul Hockman told him bluntly, "Carpenter is the only IBM management representative on AMSA."

Lt. Col R. D. Hippert, the AMSA SPO Chief, sent a letter asking for a current AMSA organization chart and any future changes.

ESC-Owego's improved business fortunes were being noticed by avionics industry publications. *Aviation Week* announced the IBM A-7 award on August 21. This was followed by a major *Electronics News* article on September 25 with the heading "IBM Leads Military Avionics Race." They described the probable sale of 1400 4-Pi Mark II computers for the F-111 and the FB-111 and more than 500 computers for the formally unannounced A-7D/E win.

The Binghamton press publicly announced the A-7D/E contract award with a page five headline, "IBM-Owego Gets $168,500,000 Corsair Computer Job." Jackson was quoted at length and noted Bill Carroll as the A-7 program manager. This four-year award, with engineering changes, spares, maintenance, test equipment and other follow-on contracts would easily exceed $200 million. Weeks liked to observe to selected audiences that these improvements in Owego's business fortunes all occurred after he became assistant general manager.

Jackson conducted the next "on notice" interview on November 10. We had a cursory review of compliance with Evans' letter and talked constructively about some of his business problems such as the Mark II Computer delivery schedule slippage. Since we had won all of the new business opportunities of the last two years, he noted that we needed to realign our technology and marketing emphasis for the longer term. Although I agreed, I noted that I could not address such matters until my current situation was improved substantially.

We talked about the A-7 Program. I advised him that my choice to remain in Owego, in spite of his actions, was because Owego senior management still owed me for past performance. He admitted that I had demonstrated no intention to become the A-7 program manager. He also characterized my August 23 call to Bob Henry not as insubordination, "just bad judgment."

He suddenly realized that he had indirectly admitted that the most serious charges against me were fallacious. He then re-emphasized that the "on notice" status would continue and added that he "would rather lose without me than win with me." I took him at his word, but forbore to congratulate him on his perfect two-year record of losing before I returned to Owego.

Since my demotion, a greatly reduced Department 566 could not be protected from further raids by former peers; another 20 percent were lost. The remaining professional staff could not fulfill the contractual obligations on the AMSA Avionics Study. This "death by a thousand cuts" had to be stopped.

Last Interview

Knowing that our diminished organization would not likely survive for long, I looked at how we might merge the AMSA and B-52 programs. AMSA would be in a holding pattern for at least another year and the B-52 enjoyed a much better near-term market development potential. The current B-52 organization had failed to exploit those possibilities for the past several years.

In my December 21 "on notice" interview with Jackson, I turned the meeting into a near-ultimatum. I noted that working for him and producing exceptional results had been very difficult and unprofitable. A substantial salary increase was called for, and if awards were ever given for the A-7 Program, that should also be considered. Then I observed that Weeks and Chow were allowing other functional managers to absorb Department 566 people in a way that undermined our ability to fulfill the contractual obligations to the AMSA SPO. Switching to a more constructive tack, I emphasized my marketable status in the avionics industry. I had managed three major proposals that had won $300 million in the past two years, practically the entire available market for new military avionics business.

Next, I proposed a merger of the AMSA and B-52 programs, which would: (1) form an organization with enough professional resources to perform on established contracts; (2) enable the restructuring of the B-52 management and staff while infusing skilled AMSA managers and professionals to improve past performance, which had been marginal; (3) enable us to apply the uncommitted collective resources to the program having the better immediate business prospects; (4) enable us to establish and pursue a broad B-52 new business initiative, such as the "B-52 Re-Instrumentation Study," by adapting the latest AMSA avionics subsystems to the B-52; and (5) enable alternating between the contract performance and proposal initiatives to minimize the latter expense, as I had done for a year on the AMSA/A-7 proposals.

Finally, if the only prospect was AMSA, Chow and Weeks, then I certainly would soon be seeking employment elsewhere.

John replied that we would get together again after the holidays. An armed truce was the most likely relationship with Jackson in the future, if any.

Chapter 7. Bomber & AWACS Advocacy, and More Avionics

Once I was no longer "on notice," I dealt with normal middle management problems of budgets, sales projections, staffing, hiring, proposals and negotiations related to the ASSB Study which provided about $3,000,000. This assured continuation of the ESC-Owego AMSA Avionics Program, but timely performance under the present circumstances was improbable.

On January 10, 1968, I told Bob Evans about my December 21 meeting with John Jackson. His involvement was essential regarding the situation in Owego and the possible merger of the AMSA and B-52 programs. I also advised him that his directive to return the IBM AMSA Program to competitive status could not be accomplished by anyone under the current muddle of Owego senior management.

On January 19, there was a series of meetings with variously Jackson, Weeks and Chow. The transfer of the B-52 Program to Department 566 was announced at ESC-Owego. We were again renamed as a function, to Strategic Aircraft Programs. This was followed by my first significant salary increase in three years. They reluctantly reinstated me as the functional manager, and they transferred Chow to a staff planning job in the Gaithersburg Space Systems Center.

All my former responsibilities were essentially restored when John Cooney, with his extensive staff and charter for AMSA and A-7D/E controls and displays, simulation and test, rejoined Department 556 a few weeks later. We were now back to about 200 people and had the B-52, A-7 and AMSA revenue to support the organization between proposals in these promising business areas. The reorganization also saved $157,500 in operating expense.

The B-52 Program

The B-52 avionics was designated the AN/ASQ-38 Bombing Navigation System (BNS). This program was the financial backbone of ESC-Owego and their principal source of revenue for a decade. During the past five years, senior management disinterest, ineptitude and preoccupation with diversity had exacerbated the predictable decline of the program. Because of poor program and middle management, and some marginal or inexperienced professionals, the B-52 BNS Program was a mess. Cost, schedule and performance problems were deeply ingrained.

I selected John Hart and Andy Coyle from the B-52 Program and Gervydas (Sim) Simiatis as the new second-level managers for the ASQ-38 Avionics System. Our most difficult performance problems were on the Low-Altitude Calibration (LAC) units being supplied to Boeing for their B-52/SRAM Modification Program. The LAC was a defective design, causing schedule slippage and a potential cost overrun. We made major shielding modifications to meet the initial flight test schedule and completely redesigned the production model. The latter effort saved the Air Force about $1,000,000 in production costs.

The Phase I & II Mod–1841 AN/ASQ-38 Terrain Avoidance Improvement Developments were completed to provide the B-52 with important new low-altitude mission capabilities. The business potential of these modifications in production quantities was about $30,000,000. The first such award to IBM was $7,000,000.

The B-52 BNS Depot Repair and Modification Program was finally completed. This contract required the repair and refurbishment of nearly 2,000 units having the old "beer can" and miniature vacuum tube technology of the 1950s. Deliveries of AN/ASQ-38 Maintenance and Overhaul Spares included 3641 units under 29 different contracts and purchase orders.

Titan III Computer

ESC-Owego's first ICBM computer was developed for the Titan II Missile. Titan III was an adaptation for launching military payloads into low-earth orbits. Department 566 inherited a serious development problem when we took over the B-52 business area.

We completed the investigation of a computer operation anomaly that occurred during a July 1967 flight at Cape Canaveral. The problem was found to be caused by an electrostatic discharge phenomenon within the computer coolant lines. FSD's work resulted in a major change to the launch vehicle coolant lines. Two subsequent Titan III flight tests in 1968 were successful, preserving IBM's potential "successful flight test incentive fee" of $96,000.

AMSA Avionics Program

Phases III and IV of the ASSB Study were completed during 1968. The Phase III Study developed a digital simulation model to determine the penetration effectiveness of a single AMSA aircraft and a total force of these bombers for a particularly severe weapon system environment. This study helped the AMSA SPO to decide on whether to configure the AMSA aircraft for a sustained low-altitude supersonic dash capability during penetration of Soviet airspace.

The ASSB Phase IV Study defined the detailed interfaces between the AMSA Avionics Subsystem and the total aircraft weapon system for two competitors, NAR and GD/FW. Boeing would not provide the necessary data.

An advanced avionics system engineering effort during Phase III and IV defined an optimum baseline configuration within the probable program cost constraints. This baseline would be necessary during the anticipated Competitive Design Phase (CDP) but allow for the later addition of equipments and capabilities, such as lethal defense.

When the government and their contractors have too much time and money at hand for a given task, the work tends to expand to fully utilize both. Due to the experience and forceful personality of Paul Hockman, the AMSA SPO made this mistake.

After developing the $2.0 billion cost in Phase II, we gradually convinced the SPO to allow phased avionics development. This spread the costs of the more ambitious configurations into later years when they would be needed to cope with the evolving effectiveness of Soviet air defenses. We allowed for the space and weight estimates needed in the aircraft for such later capabilities in ECM and lethal defense.

IBM received $7,500,000 for AMSA study and simulation activities during 1965–1968. These efforts, totaling about 200 man-years, produced an 80-volume technical data base of formal and SPO-approved technical, cost and management reports. This database and over 100 experienced AMSA professionals ensured a strong IBM competitive position for winning the AMSA Avionics Program.

The ASSB studies defined and supported the feasibility of the most complex integrated avionics subsystem ever configured. The studies also analytically substantiated the multi-mode, phased array airborne radars and ECM.

The final report for Advanced Development Task Six was submitted to the SPO in mid-August. This task represented the analysis, design and simulation of the most sophisticated and complex avionics controls and displays subsystem ever developed at ESC-Owego. Task Six had also provided the incentive and means for a major reconfiguration and instrumentation of the ESC Simulation Laboratory, now acclaimed as one of the best in the avionics industry. The SPO completed payment for Task Six on September 24, providing ESC with $2,038,500 in revenue.

The long-awaited Competitive Design Phase for the AMSA Program was announced on November 19. The CDP assured us that the Air Force was really going ahead with the prime contractor selection for the airframe and later the avionics subsystem. After six years of program development, AMSA had suddenly become a reality and the best business prospect on the horizon.

IBM, FSD and ESC senior managements responded with many high-level internal meetings, including Dr. Fubini, still IBM's chief scientist. They also depended on us to provide the substance for several government senior management briefings in the last quarter of 1968, including DDR&E and USAF Headquarters at the under secretary and assistant secretary levels.

AWACS, or Airborne Warning and Control System

With Evans as FSD president, there were frequent special studies loaded onto the mid-level managers and senior professionals. In early 1967, Evans tasked us to study the potential new missions for very large aircraft such as the commercial Boeing 747 or the Lockheed C-5 military transport plane. Given the A-7D/E preoccupation, I shelved the study until there was a little break in our excessive workloads. The study was completed in 1968 and briefings provided to carefully chosen upper management audiences in IBM, DDR&E and USAF Headquarters. Eugene Fubini, Evans and Tom Watson were involved in these briefings.

Encouraged by the results, Evans wrote to the three FSD general managers on June 14, 1968, edicting that I lead a "temporary super-group" consisting of a few highly competent professionals from the three centers to develop the AWACS systems understanding and establish a major IBM leadership role in the program.

The AWACS or Airborne Warning and Control System evolved as a Boeing 707 equipped with a huge cylindrical saucer-shaped dome, pylon-mounted atop the aircraft fuselage. The radome contained a very powerful surveillance radar antenna. At higher altitudes, this radar could detect and track moving and fixed targets for about 200 miles in all directions.

Lacking Boeing's cooperation in our study, we concentrated on the operational aspects of an AWACS. The Air Force had finally recognized that their ground-based SAGE air defense installations would be destroyed by ICBMs or SLBMs before they could be useful in a general nuclear war. This was the primary reason for developing the new system. However, the Air Force and Boeing did not yet appreciate the projected Soviet capabilities in long-range interceptor aircraft equipped with supersonic air-to-air missiles (AAMs) having effective self-guidance and ranges up to several hundred miles.

Our study showed that protection of each forward-deployed AWACS would require at least three AAM-armed interceptors with the Mach 3.0 performance of the SR-71. This possible but non-existent US capability would be prohibitively expensive. We gave Evans a briefing on our findings.

Realizing that he had a "winning hand," Evans invited the responsible Boeing vice president, Herb Stoner, and his AWACS program and technical managers to FSD Headquarters for a briefing. He also invited the irrepressible Gene Fubini. Although I personally liked and respected him, Fubini's big ego made a group meeting with him quite a chore.

Boeing was appalled and stunned by our study. Fubini was delighted and did his usual arm-waving tours around the conference room for emphasis. Bob quietly observed the patient and passive control I exercised over the proceedings. As the meeting adjourned, Evans' only comment was, "You were great."

Boeing and the Air Force soon decided that the primary mission of AWACS was not CONUS Air Defense but "tactical battle management," as it still is today. This protected the AWACS Program and allowed successful development. Now that Boeing recognized us as a serious contributor to the program, IBM eventually realized over $100 million in computer hardware and software from the AWACS Program.

The military value of the AWACS battle management system has been demonstrated in several limited conflicts, including the Bosnian and Iraqi wars where fighter Combat Air Patrols (CAPs) provided protection for AWACS. The aircraft still maintains an important, but unlikely to be needed, role in CONUS Air Defense.

Evans invited several FSD technical managers to the annual Corporate Research and Engineering Conference in Boca Raton, FL, October 22–25. During this respite they clearly identified the ascendancy of microprocessor chips and software in the commercial computer industry.

IBM Management's failure to exploit this knowledge with much larger initiatives reduced IBM's potential growth and ensured the rapid rise of two great rival companies, Intel and Microsoft.

On October 1, Mike Weeks performed my first appraisal in 41 months. Our meeting was a very neutral occasion, followed by a reasonable salary increase.

Later, Evans called to advise me that IBM corporate management would grant the equivalent of three years of salary in qualified stock options. In the late 1960s, such options were generally restricted to executive and senior management. This also implied an overdue promotion to executive manager, which Mike Weeks formally announced on January 4, 1969. The promotion and qualified stock options seemingly assured a good financial future. The several raises during the past year increased my salary by 40 percent.

Art DuBois still intended to take a respite from the endless overtime environment of the AMSA Program. More than a year had passed since his initial request to transfer, and finally he was transferred to the general manager's staff and later became manager of the technical staff. During the next 15 months he consulted frequently for us and participated in many senior-level AMSA Program briefings.

AMSA Program

There was now general certainty in the government and industry that the Air Force would fully develop the AMSA aircraft and avionics. This would be the largest strategic aircraft procurement since the B-52 was developed and produced in the 1950s. We expended considerable effort to define the costs of development, test, manufacture and support of the AMSA Avionics. In 1969–1977, we estimated the total cost would be $622,650,000. By 1974–1978, the AMSA Avionics production program was estimated at $2.203 billion.

Due to funding limitations determined by the Congress, DOD and the Air Force, initial program costs were reduced by purchasing only the most essential elements of the avionics configuration. This way, the avionics development and production costs would total about $2.0 billion in then-year dollars.

When Arthur DuBois and John Cooney were working with me on the ASSB studies, our excessive personal efforts and greater than average experience in the bomber avionics business ensured ESC-Owego a leading position in the AMSA avionics competition. IBM was understood to be the company to beat. My peers in other companies, such as Hughes Aircraft and all of the AMSA potential prime contractors, had casually mentioned their "tiger lists," lists that named the individual proposal managers in the defense industry that were most likely to win particular competitions. They allocated greater than normal resources when competing against a "tiger"; competing against IBM in the ASSB studies, NAR-Autonetics committed three times our resources. With overwork and exceptional people we always provided a more competitive performance. I was the only person from IBM on their tiger lists. IBM's competitors were well aware of my hard-earned reputation in the B-52, AMSA, Mark II, and A-7D/E programs.

Leading the pack had an important advantage. Potential subcontractors for subsystem equipments preferred to team up with an alleged front runner. This allowed us to be selective and negotiate more favorable terms for their proposal commitments, thus reducing our proposal costs.

John Jackson had a good reason for prohibiting the creation of four functional managers in Strategic Aircraft programs. IBM's salary structure placed functional managers and above on the overhead payroll; these were managers too senior to be directly charged to contracts. This move would increase ESC-Owego's overhead by $100,000 per year at the expense of profits. John also believed that none of the candidates were qualified by experience and accomplishments to be third-level managers. However, Mike Weeks failed to advise me of John's position regarding our organization, leading to greater than normal discord between us.

We organized four functions: John Cooney for the Systems Engineering Laboratory, Nick Kovalchick for Systems Engineering, Sim Simiatis for Subsystems Engineering, and John Hart for AMSA and B-52 Program Management. This organization would be necessary during the final stages of the AMSA Avionics Integration Proposal and later contract performance. Further, the new arrangement would reward many deserving professionals and create four new peers to assist me in the perpetual struggle to hire and internally recruit new professionals. We needed at least 60 immediately.

Eventually Department 566 directly employed over 200 professionals, directed and paid for 200 more in other ESC functions, and provided direction to another 200 in other aerospace companies that would supply our subsystem equipments. These professionals and three levels of management were finally organized into 39 departments in preparation for the AMSA avionic proposal.

In 1966, we had needed a good money manager for Department 566. Ernie Van Nosdall was reluctant to leave the direct path of his financial career, but he did so; and he kept us honest for five years. He became manager of our Resources Control Department in 1969. John Pfaff was my personal assistant for just about everything. As a major in the Army National Guard, he had some other demands on his time, but he managed to always be there when needed.

One of the earliest new hires was Bob Collins, who eventually became a second-level manager on the A-7D/E Program. Owego Marketing had arranged another fool's errand at ASD-Dayton and wanted Collins for technical support. Given the pressures of the A-7D/E proposals, this was refused — and just as well. As it happens, the marketing staffer in question was killed with many others in a mid-air collision approaching the Dayton Airport.

I hired two people to partially replace Arthur DuBois. Steve Wilson was a former SAC B-47 pilot, recently retired as a colonel from USAF Headquarters. Steve was very helpful in Air Force liaison and in defining the phased development of the AMSA avionics configurations in response to the evolving Soviet air defense threats. He later became manager of our department for Operational Requirements and Effectiveness.

Needing a new, all-around smart guy, I hired a PhD physicist on the personal recommendation of Dr. Edward Teller, "Father of the Hydrogen Bomb." This was like an endorsement from the Almighty. The new PhD turned out to be the worst hiring error made during my 30 years with IBM.

The AMSA SPO was confident that the next major phase of their program would be the Competitive Design Phase (CDP) to select the prime contractor for the aircraft. This would be followed by a contract award to the selected avionics integration subcontractor. Consequently, and customarily, the Air Force renamed the Advanced Manned Strategic Aircraft as the B-1 Bomber. Prime contractor selection was scheduled for December and avionics integration subcontractor for February, 1970.

This good news greatly increased the burden on the B-1 SPO, including Paul Hockman, for the B-1 Avionics. Paul sacrificed his normal crusty independence and turned to IBM for special errands. Paul was fully aware of our efforts to define the B-1 Avionics as a phased development to spread the costs. He called several times each day for a week to define a baseline avionics configuration for the B-1 competition.

As in the past four years, Paul Hockman had put out a Request for Proposal (RFP) Statement of Work (SOW) that exceeded the available funds by about $500,000. Due to B-1 Program acceleration and Paul Hockman's pushing the avionics definition well ahead of progress on the airframe, the Phase V funding was reduced to $435,000. By mid-1969, total funding to IBM on the B-1 Avionics Program was a cumulative $7,870,000 with a profit of $610,000. This profit was not the primary objective. The real benefit was a leading competitive position on a $2 billion program.

B–52 Transfer

The work, revenue and profit had progressed very well in the year that I had been responsible for B-52 Program. However, the Space Systems Center in Huntsville, AL, was falling on hard times. The Saturn V engineering and development was mostly completed. IBM and FSD senior management had made Huntsville and their revenue base a very willing captive of NASA for several years. Entrepreneurial skills and a competitive culture were neither developed nor needed.

John Jackson was tasked with transferring a major business endeavor to Huntsville. This was an easy choice for John. The B-52 had become Department 566's principal revenue base; there were no longer any pressing management or technical problems; and there were lots of former B-52 people working in Huntsville. John said to move the B-52 Program.

Only one B-52 manager voluntarily transferred to Huntsville. All of the B-52 people in Owego engineering were gradually re-assigned to the B-1 Avionics Program. This filled many of our 60 vacant positions but posed longer-term problems. With only the AMSA Program, we now had a very large organization without any fall-back program to sustain us beyond nominal Phase V revenue and proposal funding, an overhead expense. Department 566 was now in the "do or die" mode for winning the B-1 Avionics Program.

B–1 Advocacy

In mid-August a group of Democratic senators, led by McGovern and including Proxmire, Hatfield and Goodell, were trying to prevent an acceleration of the B-1 Bomber development. Largely an anti-defense group, they considered the $12 billion cost of 200 B-1s excessive and unnecessary. On August 13, they wrote to the Secretary of Defense, Melvin Laird, listing their many questions and concerns regarding the need for the B-1 bomber.

Laird referred the letter to Air Force Secretary Seamans; the Air Force action officer was Emil Block, now a lt. colonel. Emil called and asked me for a paper like the one that he had reviewed as a captain in the AMSA SPO six years earlier. I spent the next two days and nights writing, at the Key Bridge Marriott — just like a decade earlier, for Colonel D. C. Jones regarding the B-70.

Emil tuned up the resultant paper with the latest Air Force doctrines on the Triad, et al. Much of our material was later used by Secretaries Laird and Seamans in their replies and testimony to the House and Senate Armed Services committees.

When the B-1 funding issue came to a vote on September 16, the Senate rejected the McGovern initiative by a 56 to 32 vote. The senators speaking for full funding on the B-1 Program included Curtis, Thurmond, Mansfield, Smith, Cannon and Goldwater. Even the Democrats acknowledged that the unique and necessary mission of the manned strategic bomber force was bomb damage assessment (BDA) and re-strike, a mission McGovern noted had been established many years earlier (circa 1957).

During a later visit to the Pentagon, Emil Block mentioned that AF Secretary Seamans had congratulated him on "the finest paper on the manned strategic bomber that he had ever seen," and immediately promoted him. Although Emil was now a colonel, he was too young to pin on his Eagles for another six months.

I had been working very diligently on the B-1 for the last several years, and I had 50 to 200 exceptionally good helpers in this endeavor. For once, I was not a one-man show. After the premises were established in the late 1950s and early 1960s, an effective middle manager in the avionics industry could have designed a strategic bombing system capable of BDA and re-strike against hardened targets. However, the unique assistance I provided to Emil in a focused 48 hours made a significant difference to the future of the B-1, and to US strategic defenses.

B-1 Liaison & Reviews

The big market potential in B-1 Avionics caused continuous and intense senior management interest and concerns. The high-level briefings, program reviews, subcontractor selection and teaming arrangements, and the need to support the B-1 SPO seemed to be endless. Eighty and 90-hour weeks were again the norm. I kept everyone informed through weekly summary reports to the managers in Department 566 and senior management, including Bob Evans at FSD Headquarters. This practice was demanding, but had proven to be very helpful in the A-7 competition and usually prevented spurious and diverting initiatives by the many concerned parties.

On January 14, with a full back-up team, we briefed twice at SAC Headquarters in Omaha, NB. The first was a screening briefing of 20-some majors and lt. colonels. Following their endorsement, we briefed 30 colonels in the afternoon.

Major Fred Ermel knew the areas of SAC's primary interest better than we did, so he ran our viewgraph machine for the second meeting. Fred selected the most important topics in real-time without prior discussion. We never knew what would come up next, but Arthur DuBois and I had been doing these briefings for so many years that we could immediately revert to the selected topic. There were also at least a dozen SAC generals in that audience.

We had an excellent response from SAC. They invited us to come back with more detailed defensive subsystem briefings on January 27. SAC began to send regular visitors to ESC-Owego, with a particular interest in "flying" John Cooney's Two-Station (offensive and defensive) simulation facility. As our reputation spread, even some of the RAND people left their ivory tower in Santa Monica for a meeting on April 15. The visitors included long-term associates Hal Bailey, Ed Oliver, Bob Smith and others.

Tom Watson, the IBM chairman, and Bob Evans visited Owego on January 19 and I presented an hour-long overview of the B-1 Program. Evans frequently visited Owego and required periodic updates in Gaithersburg. He also arranged a review with Fubini, Evans, Jackson and Jerome Wiesner of MIT/MITRE (Massachusetts Institute of Technology).

Evans required the development of a "must win plan" and a final version of our B-1 Avionics Phased Development Plan for the pending competition. We presented these plans to Jackson and Weeks on May 10. John approved this briefing to the B-1 SPO for May 11. The phased development plan emanated from my extensive telephone discussions and May 1 visit with Paul Hockman. Paul confirmed this plan as the SPO position on B-1 Avionics development.

Mike Weeks was still a major problem, given his position as boss and ESC assistant general manager. He continually inserted himself in meetings, unilaterally arranged external commitments, and took whatever actions he could to reduce or jeopardize Department 566 control of the B-1 Avionics Program. Weeks was either not aware of or was simply inured to his limitations. Ken Driessen continually complained about Mike's activities. Whenever possible, we would accompany him on important customer or subcontractor meetings to mitigate potential damage to IBM's reputation or interests.

Weeks attempted to influence our B-1 Avionics System design by using the expertise of consultants. Fubini had resigned as IBM's chief scientist to do private consulting; he was now available to Mike. Weeks attempted to redesign the B-1 multimode radar using Fubini and Wiesner. This would undo years of careful technical work by B-1 engineers, particularly Arthur DuBois. In a private meeting with Weeks, and only two months "off notice," I made it very clear to him that such actions were both stupid and irresponsible.

But he never let up; on April 23 he arranged, with Jackson and Evans approving, a major review of the B-1 Avionics Program by Fubini, Wiesner and Dick Garwin, a senior fellow from IBM Corporate Research. Dr. Garwin is a remarkably intelligent person. After the usual introduction, Art DuBois briefed his multi-mode radar design. Fubini objected to a major feature of the radar and Arthur politely explained where Fubini was wrong, and why. Given Fubini's formidable international reputation, there was only silence. Arthur then continued his radar briefing until Fubini interrupted him, exclaiming in his heavy Italian accent, "By gosh, Art, you are right!" This exchange later became known throughout the Company and Art was often recognized as the engineer who had argued with Fubini and won.

Nick Kovalchick was the next speaker, covering systems engineering. Nick was a very solid, productive engineer who wrote excellent detailed specifications. However, he was vulnerable in the oral exchanges with these big-league critics. Fubini and Garwin soon reduced the discussion to an inquisition rather than a review. I let this go on for a few minutes and then interrupted, noting that many of the decisions they were questioning Nick about were those of the Department 566 manager, who could more readily explain our position. Further, in the interest of saving their time, the manager would give all of the remaining briefings.

The reviewers were all smarter, better educated, and better known than I was, but I had spent the last 15 years establishing the requirements for and designing strategic bombing systems. I had few peers in the business and most of them now worked in Department 566. My years of experience and overwork provided more of a lecture than a review to Mike's panel of experts. Evans said later that he had heard about the meeting — "and that you are very hard to help." After that session, B-1 Avionics briefings were understood to be progress reports rather than critiques.

Much of 1969 was spent selecting and negotiating teaming arrangements with potential subcontractors for the individual equipments required in the B-1 Avionics. IBM would supply all of the digital data processors, controls and displays. The principal subsystems included the inertial platform(s), multi-mode radar, Doppler radar, radar altimeter, infrared and electro-optical (TV) equipments, and the active and passive electronics countermeasures (ECM) hardware.

The multi-mode radar was the most complex and expensive subsystem. After an industry-wide search, we selected Hughes Aircraft in Los Angeles as having potentially the best equipment. After extensive senior management, program management, and marketing exchanges, we were in accord on everything except the digital processor for the phased-array antenna.

Hughes Aircraft considered this complex and expensive processor to be their equipment for sale to IBM. Jackson's specific direction was that we would use the 4-Pi technology to produce the radar digital processor, at a lower cost to the Air Force and greater profit for IBM. We haggled with Hughes on this issue for months. Finally recognizing IBM's strong competitive position and possible alternative radars, Hughes reluctantly agreed to our terms.

In a later meeting with Hughes in Los Angeles, Weeks insisted on contributing his usual banal philosophical generalities. Hours into the meeting, he asked, "Who manufactures the radar data processor?" The Hughes program manager replied in a loud and bitter voice, "You do!" Even Weeks was a little taken aback.

Two of the potential B-1 airframe prime contractors realized that IBM was diligently protecting their and our proprietary information during the competition: GD/FW and NAR-Los Angeles. Boeing did not share enough information to produce a detailed avionics interface specification for their design of the B-1 Aircraft. Further, they related to the B-1 SPO that their meetings with IBM were "hostile." When Jackson or FSD Marketing visited them, Boeing insisted that the IBM AMSA program manager was incompetent and should be replaced if they were ever to have IBM as their avionics integration subcontractor.

As a very well-informed member of the defense establishment, Boeing was much more aware than FSD senior management of my prior contributions to the B-52 and B-1 SPO, and my external reputation as a competitor in winning the A-7D/E Program. A couple of years later, the reasons for Boeing's denigrations would become more apparent.

Air Force senior management had become more interested in the B-1 mission and avionics. On March 24, I briefed Under Secretary McLucas with John Jackson and Paul Kossler as a technical backup. Later in the year, we briefed Air Force Secretary Seamens. He assuredly found IBM's positions on the major issues to be consistent with Colonel Emil Block's.

FSD Marketing had long (and reluctantly) recognized that a good and often private customer rapport with the Air Force that was necessary for successful new business efforts. They had also benefited very substantially from the private working relationship I had established with Bob Henry on the A-7D/E program. These relationships are based on the customer's trust and confidence in the individual representing their company. IBM customers trusted that all my actions would be in their and IBM's best common interests. After seven years of working together in the interests and support of the B-1 SPO and the Air Staff, they continued to rely on this relationship for errands such as those previously described.

Management

Mike Weeks insisted on his right to represent IBM to the B-1 SPO. His visits there with our Dayton marketing people were so tiresome and non-productive that the B-1 SPO finally refused to see him or return his telephone calls. There were similar complaints from associates on the Pentagon air staff.

With continued internal and customer complaints about Mike Weeks' assertive role in the B-1 Avionics competition, he had become an obstacle to winning the contract award. Aside from being an enduring burden and periodic threat, Weeks was now a real disadvantage to the B-1 Avionics bid. One of a successful proposal manager's responsibilities is to ethically mitigate or remove all obstacles to winning the program. It wouldn't be easy.

Mike was going to make another of his solo "goodwill" visits to General Dynamics in Fort Worth on September 25 regarding the B-1 and Mark II Computer programs. Given the many schedule and performance problems that persisted on the Mark II Program, a senior management visit without a technical backup team and solutions to GD/FW's problems would not be well received.

All of the major aerospace companies in the 1960s had huge installations of IBM mainframe computers that produced millions of "points" each month, i.e., several million dollars of high-profit revenue from each computer leasing contract. In addition to our ethical motivation for performance and fairness to the airframe prime contractors, we were always aware of IBM's much greater interest in these commercial accounts. FSD's profits were trivial by comparison. The Corporation was very intolerant of any FSD activity that might jeopardize these commercial revenues.

During our AMSA meetings with GD/FW, I had gotten along very well with John Good, manager of their avionics programs. He appreciated my forthcoming and equal consideration of the potential B-1 prime contractors. After weighing the risks, I decided that he was much more trustworthy than FSD senior management. I called him, and learned that he was aware of Weeks' intended visit. We discussed the continuing Mark II Computer problems; then I casually suggested that if Weeks' visit did not resolve his concerns, John should call the IBM Corporate Headquarters in Armonk, NY, using a number I provided. We never discussed the matter again, but I guess he made the call.

On September 26, 1969, an ESC news bulletin was posted. "John B. Jackson announced today that L.M. Weeks has resigned from the IBM Corporation to accept a position with an aerospace company." Apparently Weeks had been met at his plane, on arrival back at Binghamton, and given an hour to clear out.

Much more important management changes were brewing. In early October, I learned that Bob Evans was returning to corporate headquarters, having successfully salvaged and enlarged the Federal Systems Division. He had established the Saturn V Program in Huntsville, built the Gaithersburg Space Systems Center, and ensured ESC-Owego's future with the 4-Pi Computer Family and $300 million in new business.

As the successful ESC general manager, John Jackson replaced Evans as FSD president and Jim Bitonti took over as ESC general manager. Ken Driessen was now marketing manager for ESC-Owego. The manager of Department 566 merely remained in place to continue the heavy lifting.

Bob Evans sent me a farewell letter, the last good words to be heard from FSD senior management for many years.

Chapter 8. B–1 Avionics Competitions

Wen Chow had been transferred and the burden of Mike Weeks eliminated, but a new motif had become possible since ESC-Owego was now prosperous. The new senior management characteristics were big egos, arrogance, mediocrity and little experience in the avionics industry. The new senior managers had developed these traits in abundance in IBM's commercial computer divisions.

The new ESC-Owego general manager, Jim Bitonti, appointed Phil Stoughton to be the assistant general manager for development (engineering) and Al Zettlemoyer the AGM for operations (manufacturing). Department 566 reported to Stoughton. While I was considered for his job, it was probably just long enough for a good laugh.

I never used the phrase myself, but Bitonti very much resented being told by anyone that Department 566 had "saved Owego by winning the A-7 avionics program" before he came there. Stoughton had a similar attitude. Different backgrounds and personalities made the situation difficult and likely to worsen.

Arthur DuBois returned to the B-1 Program on January 26, 1970. He immediately resumed his overall involvement by performing an audit of the ECM subsystem status, helping John Hart on overall program costs, and supervising Nick Kovalchick's work on ASSB Phase Six.

Phil Stoughton was determined to exercise more control and involvement in Department 566 activities. He began by trying to move Graham H. Jones into the department as our manager of engineering. This possibility was vigorously refused.

Since Bitonti wanted Jones in our organization, we compromised by putting him on the staff with nebulous responsibilities for ESC products, i.e., data processors. Stoughton then insisted that Jones become manager of engineering. Arthur DuBois finally accepted responsibility for B-1 avionics engineering on April 27, with 22 departments reporting to him. Bitonti liked Arthur and his qualifications for the job were incomparably superior; this ended Stoughton's attempted manipulation of our organization.

B–1 Avionics Competition

During January, a detailed cost estimate of the B-1 Avionics competition showed $6,600,000 would be required to support Department 566 during the competition, several months of proposal evaluation, selection of the winner, and contract negotiations. The FSD Controller approved the overall plan and expenditures of $900,000 for February and March.

During the B-1 Airframe competition, we visited the B-1 SPO and USAF Headquarters to ensure the validity of the preliminary B-1 Avionics RFP and their objectives. The Air Force assured us that our preparations and phased configurations of the avionics were consistent with their intentions.

On February 23, ESC held the "kick-off" meeting for the B-1 Avionics proposal. I had been waiting and preparing for this day for the past five years. We needed all of the 39 departments in Department 566 to prepare the proposal and specifications to our 71 potential subcontractors. We were short 50 professionals, but compensated by the usual excessive personal efforts.

On June 25, I received a telephone call from a loyal associate at FSD Headquarters, warning of a Marketing move against me. Someone had written to John Jackson to complain about me; and the handwriting was familiar. The letter stated that we were behind in the competition for the Avionics Integration Role on the B-1 Program, that ESC Management could not control me, the proposal manager, and that I should be immediately removed from the B-1 Avionics Program if Boeing won the B-1 prime contract.

Our most threatening adversaries in the B-1 Avionics industry competition was no longer NAR-Autonetics, as for the past five years, but FSD and ESC Marketing, and the Boeing Company.

The final phase of the B-1 Avionics competition was to be short and intense. NAR-Autonetics and IBM were furnished draft copies of the RFP in early January. This required very expensive and extensive pre-proposal preparations for the direction and management of our selected subcontractors. We had no choice in making this pre-emptive effort, since NAR-Autonetics would win through such activities if we did not similarly prepare for the final competition.

There was a further motivation since FSD senior management had deliberately put Department 566 and its dedicated people in a "win or else" situation. My personal circumstances were even worse. Bitonti and Stoughton now constituted a second generation of hostile managers; and due to our successes in 1966–1967, they lacked John Jackson's earlier need for immediately successful entrepreneurial actions.

Our final B-1 Avionics management proposal described the B-1 program manager as an assistant general manager in ESC-Owego. Further work for Bitonti under any circumstances was unlikely; he, with Jackson's acquiescence or support, would eventually insist on having someone else in spite of the Air Force's considerable preference for working with me. The only question was "when?" Consequently, I had no motive for moderation. The pre-proposal phase was conducted as an all-out effort.

End Game

The B-1 prime contractor was selected and Secretary of Defense David Packard approved a B-1 Program initially limited to $300 million. Senator Murphy of California "leaked" the selection of NAR-Los Angeles on June 5.

From B-1 concept to contractor selection had taken eight years. During the past five years, definition of the avionics had progressed well beyond the development of the engines and airframe. Within the funding limits imposed by DOD, the Air Force decided to eliminate any work by the avionics contractors for at least a year. With the money gone, Bitonti no longer needed Department 566.

The B-1 SPO, which my voluntary work during the early 1960s had helped to create, had reluctantly done in the IBM AMSA program manager. The dissolution of Department 566 and the need to find alternative positions for the 211 employees was the worst aspect of a bad situation.

In an industry where the normal win rate was 10 to 15 percent of proposals submitted, Department 566 had never lost a bid in our five years of existence. Yet Bitonti and Stoughton stated that they had no interest in my remaining in Owego, and their transfer appraisal was such a dismal farce that I refused to discuss it.

Bitonti called a formal meeting of the 200 or more B-1 people in Owego on July 14. He thanked them for their efforts and confirmed that the B-1 Avionics Program no longer existed. After the meeting Bitonti said he was glad to be rid of the whole B-1 Program and especially me. His last statement revealed his personal animosity: "And don't believe any of this [expletive] that you were the guy that saved Owego."

By July 1, we had transferred 106 of our 211 people to acceptable new positions in ESC, FSD and Corporate divisions. The most able professionals later rose again to comparable positions.

In mid-summer 1970, I had several meetings with FSD senior managers, including Bob Evans at Corporate Headquarters, John Jackson and Craig Williamson, manager of FSD marketing. Evans suggested I remain in FSD for another year or two, due to poor external defense and commercial business conditions. Jackson said that Bitonti needed better middle management, but did not realize it. Williamson did not admit to writing the damaging letter to Jackson but said that he had discussed the matter with him. The only choice I was offered in IBM was an FSD headquarters position working for Joe Jennette as manager of market requirements.

With my tenure established, nothing would be lost if I was fired for an Open Door action. I advised Bitonti that I expected the normal six percent increase in salary for a major move, which he had thought he could avoid due to the recent farcical appraisal. Otherwise, I observed, he would have a "very interesting summer" as IBM Corporate received a detailed review of my five years of contributions and subsequent treatment by FSD senior management.

Market Requirements

The new job began on August 17 at the FSD Headquarters, then in Gaithersburg, MD. Joe Jennette was a decent man; he was now marketing manager for FSD. As the manager of the new Market Requirements Department, my first and only employee was Dr. Bill Offutt. Three or four professionals on the headquarters staff also worked with us on an ad hoc basis.

Joe Jennette occasionally asked me for peripheral studies related to ballistic missile defense, the Advanced Airborne Command Post, and hard-site ICBM basing. The primary task was Computer Aided Instruction (CAI). Commercial IBM and the local IBM Federal Systems Center (FSC) were developing a CAI console for possible government agency and public school applications. Desk-top personal computers were not foreseen and were not widely used for another 20 years.

I attended several congressional hearings and reviewed many publications regarding CAI. My November 30, 1970, summary concurred with the current FSC marketing strategy of concentrating on the rapidly emerging military market.

One of the few benefits of the new job was that it enabled me to become familiar with the workings of government and industrial institutions. Such liaison involved the Congress and "Beltway Bandits" such as MITRE, ANSER, the Brookings Institution, and the American Ordnance Association. This background was very useful in later independent work for the Air Force, DOD and Congress.

B–1 Advocacy

After five months of searching for market potential in civil programs, Joe Jennette allowed me to consider military activities. The nascent B-1 Avionics Program was still the most promising, but ESC-Owego was generally uninterested.

The B-1 SPO management regretted their role in canceling the avionics program. Just before Christmas in 1970, Jack Trenholm called to ask how I was doing after the demotion and the move to Bethesda. I advised him he could call any time if he needed help on B–1 advocacy.

Jennette agreed to a sustained and unobtrusive surveillance of the possible B-1 Avionics Program. After the B-1 was approved by DOD and the contract awarded to North American Rockwell in Los Angeles (NAR-LAD), Brigadier General Guy Townsend was replaced as the B-1 SPO Chief by Major General Douglas T. Nelson. General Nelson was much more effective with the congressional committees, but they remained skeptical regarding the need for full development and production of the B-1.

General Nelson called me on January 26 to consult on current B-1 advocacy problems. At the B-1 SPO the next day, the General explained his problem keeping the B-1 "sold" to Congress during the next six to eight months. He provided a report on the B–1 Bomber, signed by Senator George S. McGovern and Congressman John F. Seiberling for the Military Spending Committee of Members of Congress for Peace Through Law. This was an outside organization essentially dedicated to the reduction of defense expenditures by the federal government.

The foregoing document outlined the anti-B–1 arguments they planned to use in later congressional testimony. Their positions included: (1) There was little need for a new manned strategic aircraft, given America's formidable and growing ICBM and SLBM strength; (2) Including the cost of a new fleet of 255 aerial tankers showed a more realistic, and high, price tag; (3) It might be better to kill the B-1 Program for now, upgrade the B-52, and consider a Stand-Off Missile Launching Aircraft (SMLA) alternative to the B-1 penetrating bomber force; (4) It might be possible to discredit the Air Force B-1 program cost projections; (5) The B-70 cancellation provided a precedent for curtailing such a program; and (6) B-1 supersonic capabilities were too expensive.

The McGovern paper recommended: (1) B-1 supersonic capability should be dropped, regardless of other decisions; (2) Funding for the coming fiscal year (FY1972) should be limited to the costs of terminating all extant B-1 contracts and reverting the B-1 to research and development status for about $20 million in FY72; (3) An R & D program for a new SMLA with penetrating cruise missiles should be funded instead.

General Nelson was provided with the several counter-arguments that we could develop for his subsequent meetings at the House and Senate Armed Services Committees. I had known all of the technical and management principals in the B-1 SPO for many years. They would help find the necessary technical and cost data. I spent several months working with them. Jack Trenholm and I developed an overall strategy and supporting arguments for the B–1 Program.

The past 13 years had included a constructive relationship with the ASD Plans Office. The principals were now Jack Cannon, Tom Carhardt and Carl Oaks. Carl had recently completed a major study on the SMLA, F–15 and B-1; he generously provided the overall program cost data that we needed to refute the McGovern arguments.

I had two full-time jobs again; the analytical work and writing was done in Washington. Paul Hockman also asked me to come back in two weeks to help him determine the B-1 Avionics configuration and proposal ground rules for the competition that he hoped to initiate by mid-1971.

On February 11 and 12, I was in working sessions with the B-1 SPO, including Hockman, Ermel, Oaks and Carhardt. Jack Trenholm reviewed the outline of the proposed B-1 advocacy paper that would be helpful in General Nelson's congressional testimony over the next several years. Jack thought the plan to be very ambitious for one man, but agreed that this was what the SPO needed.

Liaison with the SPO gave me new hope that the B-1 Avionics Program might be initiated in mid-1971. Accordingly, I briefed John Jackson on the prospects February 17. He immediately stirred up the ESC management on this subject. ESC Marketing and Ken Driessen responded with calls to Joe Jennette, objecting to my contact with the SPO at any time for any reason.

Our Dayton FSD marketing representative, Harold Hussey, was kept generally informed regarding my work with the SPO. Harold had been in Dayton many years for FSD and as a 30-year IBM employee was a much different person than the general ilk of FSD and ESC marketing personnel. The SPO called an urgent meeting for February 23 to release an RFP to IBM and NAR-Autonetics to do an abbreviated offensive-only B-1 Avionics Study. Owego intended to exclude me from the bidders' conference, for the first time in seven years, but Harold extended an invitation.

The ASD Commander, General Stewart, gave the introductory briefing. Jack Trenholm and Major General Nelson answered questions to clarify the avionics RFP. Returning to Owego with Crenshaw and others from ESC, I continued working at the Owego Treadway Inn until 2:00 a.m. on the RFP and produced a three-page outline that described the 16 immediate tasks to be performed.

The next day Crenshaw gave a rambling summary of General Stewart's 30-minute message and intended merely to pass out the RFP and meet again on the 25th. I took over the last 20 minutes of the meeting to give the attendees a summary of the tasks that needed to be performed immediately. They reluctantly accepted the analysis. Zettlemoyer was advised of the foregoing and two days were spent getting the study initiated and staffed. There was a follow-up visit to ESC on March 1–3.

I spent March and April in Owego working on the abbreviated avionics study and alternating with advocacy work for the B-1 SPO. The IBM technical proposal was delivered to ASD on March 15, followed by the cost proposal on March 22. I performed this relatively minor exercise to help the SPO structure the costs and schedule for a minimal avionics configuration. My experience with the current Owego milieu caused me great concern regarding ESC-Owego's ability to respond to the final B-1 Avionics RFP in three months.

The initial advocacy outline I had furnished to the B-1 SPO described the usual qualitative, first-principles approach to very large problems. This included the delineation of strategic and tactical missions, the role of heavy bombers in the warfare spectrum of deterrence, prosecution, post-attack and peripheral missions. My 1960–1970 bomber advocacy papers helped to develop broad professional arguments that would be logically defensible, contribute to the public dialogue, and be useful for testing opposing positions.

As always, the B-1 bomber's role in passive and active deterrence through "show of force" and coercive warfare was described and supported. War prosecution missions included the bomber roles of BDA and re-strike, damage limiting, and assured destruction. The post-attack phase included destruction of strategic force residuals, preventing re-constitution of opposing forces, long-range interdiction missions, and support of theater operations.

Quantitative supporting arguments included the strategic target spectrum with attendant coverage and the total weapon delivery potential of the intended B-1 Bomber Force. Unique and necessary missions and several peripheral mission applications were also covered. Nine strategic mission phases and aspects were described with comparisons to the principal extant and proposed competitive strategic systems: ICBMs, SLBMs and a possible SMLA armed with cruise missiles.

Preparation of these papers and support for General Nelson's testimony to the Congress required weekly trips to Dayton in addition to my usual IBM responsibilities. My role in B-1 advocacy over many years had had very little formal acknowledgement by the Air Force, but on April 29, 1971, General Nelson sent a note praising my input in the discussion and the written information which I had prepared, concluding, "Thanks again for your help."

Jack Trenholm remembered my effort, seven years earlier, to get some favorable, or at least fair, treatment for the B-1 Program through the *Wall Street Journal*. On May 4, he suggested that we ask Bill Beecher, now with the *New York Times*, if he would consider another article on the B-1, using an unclassified version of the data produced through the recent work performed for the SPO. He set me up with a working relationship with Colonel John Thulin at the AFSC Headquarters on Andrews Air Force Base to minimize travel time.

Bill agreed to lunch at the Watergate Restaurant on May 19. I told him about my recent activities on behalf of the B-1 SPO and advised him that my work with the SPO would develop and confirm an unclassified version of the findings that he might find useful for a broader article on the B-1. The new bomber program currently enjoyed a high profile dialogue in DOD and in the public domain. Bill agreed to consider the matter, depending upon what new material could be cleared with the Air Force.

During late May and early June I prepared an unclassified paper that summarized allowable parts of the classified (Secret) studies. To ensure that there were no errors in classification, Colonel Thulin transmitted the paper from AFSC to the B-1 SPO on the secure digital communications network that had been established between major defense and nuclear weapon agencies. This network became the public Internet when Senator Al Gore "invented it" twenty years later.

The SPO was pleased with the initial effort and added some data I had thought to be too sensitive for inclusion. On June 14, Colonel Thulin advised me he had received their reply; I picked it up at AFSC and we rehashed until 1:00 a.m. At 7:30 a.m. Thulin transmitted the last version on the scrambler. The SPO approved the new copy and asked that it be provided to Bill Beecher and to both AFSC and USAF Headquarters. We met for three hours in the evening on June 17. Bill was given the unclassified paper and we hoped for the best.

On July 4, 1971, Bill Beecher's article on page 1 of the *New York Times* reported his usual broad and unbiased treatment of the arguments offered by several DOD officials, six senators and the Nixon Administration. The latter had included $370 million in their budget request to develop and build three prototypes of the B-1 Bomber. The production decision was to be deferred until extensive flight tests of the prototypes proved the intended B-1 performance in range, payload and operability.

Bill reported some of the new data regarding displacement costs for operating 255 B-52 G & H Models and the necessary KC–135 Tankers. The operational characteristics of a much smaller B-1 and KC–135 force, capable of comparable effectiveness, would provide enough annual savings in ten years to pay for the entire B-1 production program. Further, these projections did not include the cost of necessary major refurbishment of the entire B-52 fleet during the 1980s.

Major General Nelson, et al., prevailed in their arguments to fund the B-1 prototype development. Colonel Thulin presented me with a scale model of the B-1 that has had a place in our homes for 35 years.

B–1 Avionics Finale

Since my enforced departure from ESC-Owego, the new senior management attitudes were reflected in the choice and character of several new middle managers. This deterioration had a price — in this case, about $2,000,000,000.

ESC-Owego asked me to visit NAR-LAD on March 15, 1971. Dick Walker, a NAR-LAD vice president, was introduced to Sam Pulford and Jim Crenshaw. We had a useful discussion on B-1 Avionics prospects. A similar errand was completed at Hughes Aircraft.

B-1 Program activities intensified during May. The SPO asked for continued advocacy liaison with Colonel John Thulin at AFSC and Colonels Traphold and Jasper Welch on the Air Staff.

A newly anointed middle manager, Lee Grebe, visited the B-1 SPO in May. John Jackson visited General Nelson on June 19. Trenholm suggested I make a similar visit to improve the prospects for writing IBM's avionics proposal because he was concerned about losing the quality of our previous work. The visit was non-substantive and duly noted at ESC-Owego.

Paul Hockman wanted a last discussion of his proposed B-1 Avionics Program. On July 15, we talked all evening about most aspects of his intentions. He expected his Request for Proposal (RFP) to be released by mid-August.

In a joint meeting with Joe Jennette and John Jackson on July 22 I summarized my work for the B-1 SPO and Paul Hockman. I asked to be the proposal manager, as I was the most likely to win the avionics program. This request was only to write the proposal and ensure that we would utilize, protect and benefit from IBM's 300 man-year effort in the B-1 Program. I certainly had no desire to work in Owego again. I was surprised and severely disappointed when Jennette recommended to Jackson that I should have only an advisory role in the proposal effort.

Jackson immediately called Bitonti and told us to charter a flight to Owego for a 5:00 p.m. meeting. Bitonti and Zettlemoyer were briefed for an hour. They met alone for another hour. At dinner, Zettlemoyer agreed on my assignment as proposal manager and, if IBM won, a role monitoring the B-1 Avionics Program and ensuring performance from FSD Headquarters.

On Friday, July 13, Zettlemoyer allegedly informed Lee Grebe that I was the new proposal manager. I discussed assignments for the pre-proposal and proposal efforts with several former Department 566 managers. Arthur DuBois, John Cooney and John Hart had been decreed "unavailable."

Saturday morning began an intense and lonely professional and intellectual effort to produce the pre-proposal and proposal plans for the B-1 Avionics Development Program. I was done by 2:00 a.m. on Sunday. This would have required several weeks of effort by a group of professionals, especially in the new ESC-Owego. The report included: (1) the Air Force Program Management Plan and the role of the B-1 Associate Avionics Contractor; (2) B-1 Production Avionics Subsystem Configuration and selected equipments/providers, (3) B-1 Production Avionics Subsystem Preliminary Development Schedule for November 1971 through April 1975; (4) B-1 Avionics Pre-proposal for July 24 to August 22 and the proposal preparation schedule for August 20 to September 20; (5) Group Leader candidates for the eight major aspects of the proposals and their specific responsibilities; (6) ten general requirements for the Group Leaders with their page allocations for a 100-page-limited proposal.

I gave it to Zettlemoyer and Grebe at 8:30 a.m. During the following two days of meetings with the selected Group Leaders we made good progress on Statements of Work (SOWs) and subsystem specifications. Zettlemoyer later said that he had not told Grebe who would be the proposal manager and suggested I return home for the weekend.

On Thursday, July 29, Bitonti assured Grebe that he was responsible for the B-1 Avionics Proposal. Faced with this collective and calculated deception by ESC senior managers to take advantage of my established initiative and expertise, I left. On August 1, Joe Jennette called and told me not to go back to Owego; Zettlemoyer confirmed that message. My former secretary, not knowing the background, called Grebe's office. She was advised that "Mr. Carpenter has been thrown out of the facility, never to return."

Paul Hockman confirmed to me the dollar amounts and allocations of intended B-1 Avionics funding that we had earlier determined to be appropriate. This information would gradually reach industry marketers; IBM had to have these numbers. Since John Jackson was going to Owego, I gave the data only to him. This gesture to help Owego, in spite of their devious and unprincipled behavior, eased the tensions somewhat. On August 6, Jackson agreed that the B-1 Avionics Program was a certainty, and added that he was satisfied with Owego's proposal work for the B-1 competition. He had already agreed that Grebe should run the B-1 Avionics Proposal. John said I should consider a lesser role in the effort and talk with Bitonti.

Bitonti said that there had been a "misunderstanding" and I should continue work on the proposal in Owego in some capacity. Owego management would be hard to help; the past two weeks had proven that they were not worth helping. I needed some other motivation to work in Owego for the next several months of proposal preparation and negotiations. Bitonti said I should establish a working arrangement with Grebe. He also agreed that the B-1 Avionics Program would be monitored from the FSD Headquarters to ensure adequate performance.

At least, I wanted to protect IBM's business interest in the B-1 Avionics Program that I had worked to establish over the past decade. Further, I intended to bring many of the former Department 566 managers and technical professionals into the proposal preparation and specifically write them into the proposal documents. Their five years of arduous and excellent work had earned them much better jobs than those they currently enjoyed.

John Cooney came in to work on his inherited F–14 displays problem, a $2 million overrun with no hope of resolution. John had been offered the job of B-1 avionics program manager as an inducement to work on the program, but the stress imposed on him by the new Owego management milieu caused him to refuse greater responsibilities. With him in the job and me as his helper, IBM would have had a much better chance of winning the program.

FSD senior management explored teaming arrangements to strengthen their business position. Had they exploited the advantages my organization established for them during 1965–1970, reinforced by the more recent work for the B-1 SPO, these companies would have approached IBM for such affiliations.

On August 8, I prepared a position paper that considered the merits and problems of teaming with each of our competitors. Due to the B–1 prime contract award to NAR-LAD, our five-year nemesis, the NAR Autonetics Division was no longer a major competitor.

The principal competitors for the B-1 Avionics Program were now IBM-Owego, Hughes Aircraft, Raytheon-Bedford, General Dynamics-Fort Worth and Boeing-Seattle. The latter two airframe companies had large and capable avionics departments to support their manufacture of military aircraft. The position paper rejected teaming and encouraged strong subcontractor affiliations to mitigate FSD technical weaknesses, such as the Stores Management System (SMS) and electro-optical sensors (TV and Infrared).

Boeing's long-term backup strategy during the late 1960s competition for the B-1 Airframe had always been to compete for the avionics if they lost their B-1 prime contractor bid. Given my established positions on the industry's tiger lists, it should have been obvious why Boeing had been denigrating my professional abilities as the IBM ASSB study manager during five years of the B-1 Avionics Study.

Boeing was also aware of my many years of independent work for the Air Force, B-1 SPO and my competitive record on the A-7D/E and B-1 avionics. Boeing had every intention of eliminating such a formidable participant from the B-1 Avionics competition. They should have spared themselves the effort. Jim Bitonti's big ego and some lesser connivers did it for them.

The pre-proposal effort was very understaffed. The first of many meetings on the avionics computer configuration with Grebe, Vandling and Kovalchick was too confused for further participation. I just added a separate "Supplement I" to the guidance document of July 25.

A primary B-1 computer design was recommended that contained two central processor units (CPUs), large enough to also control the active and passive electronics countermeasures (ECM) equipments. The data processors would be modified F–15 computers in a multi-computer configuration capable of growth to a more complex (and expensive) multiprocessor.

The aircraft electronics data bus (E-MUX) and a separate but compatible avionics data bus (A-MUX) were described. The latter was our responsibility as part of the computer complex. Another major item included the rotating memory drum(s) for display buffer storage, et al.

Other topics in the supplemental guidance included: (1) equipment dispositions and modifications for the three Production Avionics Subsystems; (2) displays for the Electronics Officers, and (3) electronics integration laboratories.

On August 10, Grebe distributed his new proposal schedule, completely ignoring all of my guidance documents. At 10:00 p.m., Paul Hockman called to again to verify the schedule previously provided.

Tom Smith, a long-time and able writer in technical publications, was asked to find the hundred or more stored B-1 classified documents that we had produced during 1965–1970. The 300 man-years of work reported in these reports would be very helpful in writing the proposal. He did find them, but Zettlemoyer and Grebe refused to release the documents to the proposal writers. They apparently wanted any subsequent win on the B-1 to be independent of earlier efforts.

On August 12, I submitted "Supplement II" to describe a new schedule for August–October 1971 proposal preparation since the B-1 SPO would not be able to release the RFP until August 30.

Supplement II also reasserted the need for a "top-down" program cost estimate by August 27. Lack of applied manpower precluded the more important "bottom-up" estimates that would finally be needed for responsible and competitive pricing. The recipients were also reminded of the ABCs of the logical and necessary evolution of the proposal sequence: (1) basic equipment configuration; (2) program performance interval that included the manufacturing plan; (3) program controls consistent with scope and funding, and (4) total program costs.

Most of these points had been made 18 days earlier but nothing was forthcoming from them, and a preliminary plan was needed by August 23.

Joe Jennette came to Owego for his first visit regarding the proposal on August 12. I gave him a written critique on the lack of proposal progress. Jennette, as FSD Director of Marketing, sent some of the proposal critique to Zettlemoyer. On August 19, I went over Jennette's memorandum to Zettlemoyer with him and others. They considered the summary to be very critical, but they became more constructive in their responses. As a first step, they assigned Bob Orrange to re-write the program management proposal. Grebe later agreed that he would work with Simiatis and Kovalchick. I would work with the other group that included Tom McDonald on the Simulation Laboratory and Jim Kiernan on the SMS and E-0 sensor problems.

I visited the Fairchild Company near Germantown, MD, with several Owego people to discuss Stores Management System (SMS) technology. Our earlier studies did not cover wing pylons to carry tactical weaponry. This was a new DOD requirement to place more emphasis on tactical warfare.

August 27 was the first all-day top-down cost estimate meeting. In the absence of any of the several proposal managers, I ran the meeting by default. We tried to reach a "bogey" with a low-quality system configuration. The results were not good, but were at least a start.

Calls to the Air Force determined that the RFP might be delayed again. John Foster in DDR&E was objecting to the B-1 Avionics Program structure, but he was later over-ruled by Secretary Packard.

Zettlemoyer's next request was for a summary evaluation of Owego's pre-proposal effort. On August 31, these were our limitations: (1) We had a non-competitive and potentially non-responsive disposition of program costs due to excessive program management, understated hardware costs for the computer complex, and inadequate consideration of the system integration laboratories; (2) About 70 percent of the program costs warranted critical review; (3) The proposed minimum and non-responsive computer complex had inadequate memory, ill-defined memory drum design, specifications and costs; the data adapter definition was still formative; and it had a non-competitive posture for growth to a multi-processor. This negative litany continued regarding system engineering, the integration laboratories, SMS and E-0 sensor procurement, and the overall program management plan.

Probable delays would allow for corrective measures, so I added "Supplement III" and gave Bitonti and Zettlemoyer a list of 15 managers and engineers from the former Department 566 organization who could be assigned to the thus far inadequate pre-proposal efforts.

Joe Jennette called regarding probable delays in the RFP release and wanted me to spend more time with him on non-B–1 tasks. Given the effort I had expended versus the results achieved, this was a welcome relief. I had given ESC-Owego all of the information and guidance that they would need — they simply had to stop vacillating and get to work.

During September, I made three more trips to Owego to monitor and encourage progress on the pre-proposal efforts — with little success. Owego was spending $25,000 each week with very little to show for it.

I let Zettlemoyer know that the pre-proposal should now be at least 80 percent complete, but there was nothing new of real substance. Further, the key people I had identified had not been assigned to the B-1 effort. Grebe still intended to submit a computer configuration that was non-responsive to the proposal definition, and he had spent $242,000 in the past two months. With the RFP at last imminent, our pre-emptive preparations were less than one-third completed.

On September 28, 1971, Harold Hussey chartered a flight from Dayton and delivered the 23-volume B-1 Avionics RFP. We had been working and waiting for eight years. A quick review of the RFP ascertained that all of the guidance that I had provided to Owego since July was correct. In a large measure, the pre-proposal response was non-existent or inept. Jim Bitonti was advised of the new urgency in our proposal response and finally agreed to transfer 12 of the former B-1 managers and engineers to the proposal effort.

On October 1, during an otherwise directionless meeting with Grebe, Crenshaw and Orrange, I listed the 23 required proposal documents along with the assignees available or required. The heterogeneous leadership required a call to Bitonti to insist that he designate a technical proposal manager. He selected Crenshaw.

The formal computer RFP from the Air Force would finally force Grebe to stop his non-responsive second-guessing on the computer complex. I spent Friday night analyzing the SPO's computer requirements. On Saturday morning, we met yet again. Sunday, Jim Kiernan came in and we worked on SMS problems until Orrange and Crenshaw arrived at 1:30 p.m. We also worked for several hours on the system specification.

Monday morning I advised Bitonti that if he did not get the proposal moving before Friday, it could not be saved. He was losing the B-1 Avionics Program. The same message was repeated with Zettlemoyer.

Grebe did not want John Cooney assigned to the proposal due to his superior credentials. When John was immediately assigned, Grebe said he was "not up to speed." I told him that John had been up to speed for the last six years.

Zettlemoyer was again asked to select a B-1 program manager from their choices of Grebe, John Cooney and Del Babb. John did not want the job and Babb eventually refused. Other topics included a proposal funding increase from $217,000 to $700,000, specific professionals and managers needed on the proposal, and identification of critical proposal tasks.

This late in the proposal, my objective was to keep IBM from being disqualified for non-responsiveness. The managers and professionals from our former organization such as Kovalchick, Kiernan, Cooney, Simiatis, Hardie and Larry Cooper were accustomed to the long hours and weekends required in major proposal efforts. The proposal manager(s), Grebe, Crenshaw and Orrange, were much less evident after normal working hours. Nick Kovalchick became particularly discouraged. I worked closely with Nick because he was seriously over-tasked, under-staffed — and under Crenshaw.

Bob Orrange, who was responsible for the B-1 management proposal, was over burdened with the task and was receiving directions from too many of his managers. I provided him with a nine-page listing of the SPO's specific management requirements for his proposal, thus removing these non-productive and divisive pressures.

My guidance to Owego on the B-1 computer complex on July 25 was definitive, but it wasn't followed. This was confirmed during the earlier and thorough analysis of the computer RFP. As usual, the SPO asked for more than was immediately necessary and well beyond the expected allocation of funds. Lack of clear leadership on the proposal effort turned this typical problem into a major dilemma that had lasted for three months. Time was running out.

On October 31, I sent a five-page memorandum to Crenshaw and Vandling (Computer Design) on the Computer Complex Proposal summarizing the essential contents and recommended configuration(s). This summary was finally accepted, perhaps by default, as definitive for the computer-related portions of our proposal.

Total System Performance Responsibility (TSPR) was required of the B-1 prime contractor, NAR-LAD. Since the selected avionics contractor would be an important part of NAR-LAD's TSPR obligations, all of the competitors for the avionics were required to negotiate a clearly-defined interface with the prime contractor in support of the latter's TSPR. Much time was spent in October working with Jim Sharp in ESC-Contracts to propose an acceptable Associate Contractor Agreement (ACA).

I provided a summary of the principal problems and recommendations on October 18 on the major aspects of our NAR-LAD contractual interface. These included: (1) weapon delivery accuracy for both SRAM missiles and gravity bombs along with BDA performance; (2) delineation of responsibilities with other associate contractors for ECM and E/O sensors, and (3) general avionics system facts such as weight, volume, power, reliability and maintainability requirements. Our Phase II and III ASSB studies, conducted during 1966–1968, had carefully delineated these shared responsibilities. By edict, these reports were still unavailable to the proposal team.

ESC Management at all levels was too slow to recognize the many limitations on pricing proposals for bidding on such a large contract, especially two primary concerns: total program cost and the time-phasing of expenditures over a forty-month interval, consistent with the funds expected to be available from the Congress.

Owego's original total cost estimate was over 40 percent too high and the progressive and cumulative expenditures were too front-end loaded for the sequenced funds available. We worked with Jack Fanning in the ESC-Controller's office and all of the principal proposal managers to solve these problems, but with very limited success. The general sense of well-being established at ESC-Owego during the Jackson tenure made them all very risk averse.

Our 27-volume proposal was much too large for a detailed review. A prioritized list was provided on October 27 for those volumes most critical in the competition. Much of the remaining time was spent on literal responses to the RFP to avoid disqualification. The proposal production process was much too large for ESC-Publications. Owego temporarily moved entire departments to a Philadelphia printing firm to ensure meeting the proposal deadlines.

My success in winning the B-1 Avionics Program made me the de facto FSD Headquarters proposal representative. There was no time for the normal Headquarters DCA approval process; I briefed the 15 principals there on October 28. They chartered a flight to Owego for the final DCA meeting on November 3. At that point, we still had four major problems. (1) Progressive funding required that we limit spending to $22.4 million during the first 16 months with the attendant schedule and manpower phasing risks. (2) Total program cost must be much less than the currently proposed $90 million. (3) We would have to bid two computer complex alternatives to be both responsive and competitive. (4) The two-week NAR-LAD Associate Contractor Agreement had to be negotiated prior to the award. And (5) total proposal costs had to be considered for a program similar in scope to that envisioned in 1969–1970, but with one fourth the immediate revenue potential, six months of pre-proposal/proposal effort and expense, and peak manpower requirements of more than 200 people.

End Game

I addressed every aspect of the proposal — technical, management and cost — wherever there was a need. Crenshaw, Grebe, Marketing and Caldwell cooperated only when they had no choice, but Zettlemoyer had actually become supportive of most of these efforts. Orrange, John Stalma (displays) and of course the former managers in Department 566 were all very receptive to this assistance. Jerry Gwinn, a long-time friend and former B-1 professional, remarked that the last three months had certainly changed my image in Owego. He said, "You would likely work with the Devil himself to win the B-1 Avionics for Owego."

The final DCA was held on November 4. Grebe casually mentioned $70 million as the final program cost objective (without giving a clue on how we could responsibly achieve that goal).

The cost proposals on large program competitions are generally submitted two or three weeks after the technical and management proposals due to their complexity and pre-determination by the earlier documents. On November 6, at Zettlemoyer's request, I prepared a summary describing what could be done legally and credibly to submit more competitive cost proposals.

In their first estimates, ESC-Owego would have proposed a selling price of $50.8 million for the initial 17 months of the B-1 Avionics Program. This was $28.4 million over the available funds for the first half of the performance interval. The program structure had already been deferred to the maximum believable extent, but if we delayed program start-up by one to several months, we might be able to reduce front-end costs. I asked Lothar Hermann and Jim Sharp to rework ESC and subcontractor prices to reduce mid-term costs by $8.5 million and hopefully much more; and re-examined and literally interpreted the RFP to determine the legal extent of cost constraints.

We knew that the B-1 SPO would indicate their own preferences for all of the avionics subsystems, especially the stellar-inertial platform, forward-looking infrared sensors, and low-light-level TV. This could be used as an opportunity to renegotiate all of our subcontractor costs and progressive payments. This could be legally done and would reduce our total program costs by another estimated $6.0 million.

The next day Zettlemoyer finally assigned the Controls and Displays Cost Proposal, covering interrelated computer complex and facility functional support, to Simiatis. It would require 214 man-years of design and development efforts.

With the final cost proposal volume provided to ESC Technical Publications, the proposal was at last completed at midnight on November 13. I had lost ten pounds during this Owego stint.

The final FSD Headquarters DCA was held from 2:00 to 4:00 p.m. John Jackson approved a final total price of $69 million for the B-1 Avionics Program. Harold called at 5:00 p.m. from the B-1 SPO to say that he had beaten the deadline by 15 minutes.

All major competitive proposals have a very critical period of several months when the sponsoring agency submits new general questions and many specific questions to each bidder. Such questions, when thoughtfully and promptly answered, can be used to strengthen your position or correct existing proposal deficiencies — but every response must be immediate, thorough and constructive. ESC-Owego simply was not up to the task.

Then the SPO made things even worse. Using our unsolicited auxiliary proposal as a basis — the one that was not directly responsive to their specifications yet was adequate and very competitive — they required that all bidders re-compete the computer complex and resubmit all related technical and cost proposals. All of our competitors now revised their bids to our configuration, thus eliminating IBM's advantage from this initiative. The new computer complex bidders' conference was called on January 22, 1972.

The many interim trips to Owego included the pre-award ACA negotiation with NAR-LAD required by the RFP. I participated due to my longstanding good relationship with NAR-LAD. During the two-day negotiations, my calls to the B-1 SPO clarified five items in contention on which IBM finally prevailed. One member of the NAA party said they were "aware of Boeing's efforts to remove you from the B-1 Avionics competition."

Zettlemoyer requested my more direct involvement in the post-proposal responses. ESC efforts were simply too disorganized to produce high quality and timely responses to the many B-1 SPO inquiries.

A four-day negotiation meeting was called by the SPO during one of my trips to Owego. I went to Dayton with the IBM negotiators — over Grebe's objections. The Owego negotiating team included ten people, but the exchanges became so demanding that another plane-load of technical and secretarial people flew in on January 5, 1972. Paul Hockman was necessarily circumspect, but he did let me know that IBM was technically last on the raw scores.

On March 1, I was asked to help answer 20 new questions from the SPO. Crenshaw and Grebe were on vacation. The temperature in Owego dropped from 56 to 41 degrees in one minute. A Mohawk airliner crashed in Albany, killing 17 people, but I went anyway.

The SPO had requested a cost-benefit analysis on certain aspects of our proposal. The initial work was scrapped and a new effort started on the controls and displays. The task was completed in three days. On April 3, after attending two more negotiations in Dayton, I met with Jackson and Jennette on the need to hold to the lowest possible total program cost and provide FSD senior management involvement. With all of the post-proposal activity, our cost had increased by $10 million.

Joe Jennette called on April 13 to say that IBM had lost the B-1 Avionics Program to Boeing. We ended the competition in second place on the technical aspects; and they finally underbid us by $15 million.

It became clear that Boeing had just plain beaten us in every respect. Paul Hockman said privately that a major factor in the IBM loss was the personal and technical arrogance of the IBM negotiators and our weak responses during the post-proposal period. The SPO simply did not want to spend the next decade working with such a non-responsive organization. Trenholm privately re-iterated Hockman's assessment and then, noting that IBM was now completely out of the B-1 Program, he offered me a permanent GS–15 position at ASD-Dayton with the responsibility for B-1 advocacy. Although I was very tempted by Trenholm's offer, I had re-located my family three times during the past eight years because of B-52, B-70 and B-1 problems and opportunities. Further, the past decade on the B-1 Avionics and advocacy had been very demanding and an extended respite was needed. Jack also agreed that B-1 Program progress was now assured for another two years, until the production decision.

Chapter 9. Program Analyses & Selection

Space Shuttle

The contract to build the Instrument Unit or fifth stage of the Saturn V Apollo Launch Vehicle was nearing completion in early 1972. Art Cooper and Ed Smythe had transferred back to ESC-Owego to prepare for a bid on the new Space Shuttle program.

The Space Shuttle configuration was determined by several subjective, circumstantial and political factors. The Apollo Program had developed a NASA staff of about 17,000 people, primarily oriented toward manned space flight. NASA wanted to sustain this capability. Consequently, the new major space vehicle had to be manned. Further, manned space flight was more appealing to the media and the public.

The original concept was based on a recoverable and re-usable spacecraft, about the size of a Boeing 707, riding piggy-back on the huge new C-5 Military Transport, before separation at high altitude. The C-5 was essentially the re-usable booster. Technical problems delayed C-5 production. NASA was forced to consider a smaller spacecraft, about the size of a DC-9 commercial aircraft, powered by both liquid and solid rocket boosters.

NASA required that the new space vehicle be reusable. This warranted two very large, solid-propellant boosters flanking the main liquid fuel tank. After the initial boost phase of the Shuttle launch, the solid booster casings were recovered from the ocean to be refurbished and re-used. The Challenger and Columbia Shuttle tragedies were related to this booster configuration.

The Shuttle spacecraft included the main liquid-fueled engines as part of the Shuttle. Fuel from the huge centerline-mounted tank was routed through the Shuttle engines for launching and acceleration to orbital speed.

The Soviet Union eventually copied the NASA Shuttle vehicle rivet for rivet, with one major exception — the main engines remained on the booster, not within their spacecraft. The Soviets made one unmanned Shuttle flight test before the collapse of their government and economy.

NASA wanted the near-earth orbital missions to be considered as frequent and routine. Hence the name "Shuttle," borrowed from the Eastern Airlines shuttle service between Washington, New York and Boston. Shuttle missions were ambitiously projected toward 20 flights per year.

Jack Trenholm, drawing on his Dynasoar Program experience, noted the Shuttle funding was insufficient for a small fighter aircraft. Spacecraft development delays were inevitable as the Shuttle Program progressed enough to merit additional, annual congressional funding.

Many scientific Shuttle missions could have been performed by much less expensive, unmanned and expendable spacecraft. We could have deferred the manned missions to a later generation of spacecraft. Successful operation of the current Shuttle configuration is a great tribute to our scientists and engineers — but especially to the courage and skills of our astronauts.

Just because an expensive endeavor is unique, awesome and technically possible does not necessarily mean that it should be undertaken — as shown by the B-70 cancellation in 1959 and economic burden of the European supersonic transport. Any great state should spend two or three percent of their annual budget to sponsor and direct major scientific endeavors for the ultimate benefit of their national entities and mankind. However, we should give more thought to "what" and "why" and only then to "how."

SECRAC

The Army had been developing the SAM-D ground-based anti-aircraft missile for several years, but with excessive program costs. The Army initiated the System Engineering and Cost Reduction Associate Contractor (SECRAC) Program to assist their prime contractor, Raytheon, in reducing development and manufacturing costs.

IBM-Huntsville asked for assistance in preparing their proposal, due to my prior work on the Nike Program. I met on February 14, 1972 with Sam Pulford, Bill McLain and Gordon Dolittle. The latter was their intended SECRAC proposal and program manager. He soon proved to be very similar in ability and attitude to some of Owego's lesser middle managers.

This was the first FSD-Huntsville bid for an Army program — after ten years of co-location with the Redstone Arsenal. I contacted an old friend, Ed Rosenfeld, still toiling for the Army Corps of Engineers at Redstone, to obtain some subjective background on the SAM-D Program. Hughes Aircraft (HAC) was the leading contender for the SECRAC Program. Ed privately advised me that the Army would delay their SECRAC RFP. FSD had not foreseen this possibility.

FSD-Huntsville made negligible progress with the pre-proposal effort from February to April. With the RFP release imminent, I went down again on April 22 and wrote a new management proposal over the weekend. After reviewing it with their proposal staff, I gave a tutorial on writing proposals, starting with analyzing the RFP line by line and writing an acceptable response for each item. Most of the proposal staff welcomed the information.

On April 27, Doolittle called to say that I was no longer needed. On May 12, he called again, to ask for help with questions from the Arsenal on their SECRAC Proposal and preparations for possible contract negotiations. I rewrote the worst of his technical proposal. Since IBM's competitive prospects had been improved substantially, Dolittle intended to give a personal presentation during the negotiations on the ABCs of surface-to-air missiles. This was neither necessary nor appropriate; his briefing would only demonstrate his arrogance and technical weakness. When he persisted, I prevailed with Sarahan to preclude such a fiasco.

Sam Pulford and Dolittle did not want me at the SECRAC contract negotiations. After ensuring that Bill McLain would be involved in the negotiations and that Larry Sarahan agreed to make some supportive high-level calls at the Arsenal, it was a pleasure to go home on May 14.

Pulford called on May 19 to advise that they had won the $2 million SECRAC contact. The follow-on contracts provided $10 million in revenue before the Huntsville facility was finally closed.

I had a long, frank discussion with John Jackson on June 9 regarding the real reasons for the B-1 Avionics loss and the SECRAC win. Management and professional arrogance and mediocrity had affected FSD business prospects and internal morale. Two days later Jackson held an unusual meeting with FSD directors and other senior managers, and according to Jennette, he emphasized many of the concerns aired in our prior meeting regarding FSD's status and future.

A closet intellectual and futurist needs to rethink his field of interest every few years to integrate new and foreseeable realities. I needed to find a major professional assignment that would require and support such activity.

The Federal Systems Division was doing too much marketing with too little thought. The work of selecting, evaluating and directing current and anticipated business endeavors requires objectivity and detachment.

Joe Jennette was gradually convinced of the need to establish and formalize the foregoing process as a normal senior management responsibility. He agreed to make this my primary assignment, in addition to specific study requirements from FSD and Corporate managements.

Higher Defense Costs

One of the peripheral tasks Joe assigned required analyzing the causes of higher defense costs. Inflation during FY1939 to FY1973 had decreased the purchasing power of the dollar by more than 400 percent. A P-51 Fighter purchased in 1945 for $54,000 would now cost $172,000. Inflation accounted for 87.4 percent of the increase in defense spending between FY1961 and FY1973.

Social pressures also increased defense costs by eliminating the military draft and raising the base pay of recruits from $21.00 per month in 1939 to $332.10 per month in 1973. The apparent increase of 1580 percent was reduced by inflation to a real increase of 379 percent or the equivalent of $79.50 per month in 1939. Real pay, allowances and retirement had also increased by 28.6 percent. All forms of compensation for military and DOD civilian employees accounted for $42.8 billion or 56 percent of the 1973 defense budget.

Advances in technology had allowed a $10.5 billion cost reduction in real acquisition, RTD&E and construction costs to offset the foregoing social costs. Fewer, high cost weapon systems having much greater performance and effectiveness allowed the United States to reduce overall defense spending while ensuring formidable deterrence to the ever-more-powerful Soviet Union.

A comparison of the 1945 P-51 fighter-bomber and the 1973 F-4 aircraft performing the same missions exemplified the foregoing trends of higher performance and fewer numbers of aircraft. The F-4 was essentially twice as big as the P-51 and had five times the maximum weight. However, the F-4 turbojets produced 40 times the effective power of the P-51 piston engine. The size and power of the F-4 afforded these advantages: 3.6 times the maximum speed, 1.7 times the altitude, eight times the payload, 23 percent greater range and ten times the rate of climb. These F-4 capabilities allowed ordnance delivery at 29 times the ton-mile rate of the P-51. All this came at a price; the F-4 cost $2,562,000, about 15 times greater than the comparable P-51 cost.

Clearly, FSD's business future depended on the careful selection and diligent pursuit of the few major RTD&E programs that promised a significant, but smaller, production potential. I had spent a decade doing this for them on the B-1 Bomber — and they blew it.

World Trends and Program Selectivity

My six-month effort produced two principal results. The first was the development of a selection process for major business endeavors for use by senior and middle managements. The FSD population had been reduced from 13,500 to 12,000 people in response to IBM Corporate direction. Careful program selection was critical to supporting this population, particularly through 1973–1975. I offered a single-page Evaluation Criteria for Major Business Areas/Programs with three columns to assign strategic value, current importance, and relative merit. Using this, FSD directors evaluated eight programs or business areas: Post Office, Education, Anti-Submarine Warfare (ASW), Medical, Space Shuttle, Trident (SLBM submarines), Safeguard (ABM) and SAM-D SECRAC.

The directors rated all the programs at an average of 69 percent acceptable and their staffers, who were closer to the attendant problems, rated all programs at 66.4 percent. Marketing people favored almost everything. The lawyers were leery of potential FSD/IBM corporate charter conflicts; the financial people favored big, immediate high-revenue programs; and the technical people were generally more cautious — they were the ones who understood the difficulties of competitive design, performance and winning.

Both management and staff gave preference to Trident, ASW and the Post Office. The Space Shuttle was in the middle at 69.2 percent. Safeguard, SECRAC and Education were all ten or more points lower.

Having consulted with many senior professionals, including Bill Offutt and Colonel Jack Henderson, over a period of months, I also provided a new fundamental view of the current and prospective global military and business environment: political, economic and military topics covering 19 factors that would influence the future of FSD (and IBM). Domestic trends covered six aspects of the next decade in social and political evolution. A classified supplement was prepared for senior management to further justify the positions taken. In retrospect, this paper correctly predicted economic warfare in the form of the Middle East oil crisis and the selective modernization of tactical and strategic forces. Progress by the new European Union in becoming a world power center was over-estimated.

With the stalemate-by-design for NATO and the Warsaw Pact nations, and Mutual Assured Destruction established between the US and USSR, the greatest danger of a general war at the time was on the Sino-USSR border. The Soviets had dedicated a portion of their nuclear arsenal to such a possible war and had 49 armored and infantry divisions on that border.

The overall reception was favorable, although some of the FSD entrepreneurial managers had established preferences. The FSD senior management generally appreciated the work, but Joe Jennette insisted that the policy implementation task for FSD Selectivity be taken over by Abe Katz's FSD Plans Department. Abe's group could not cope with the Center senior managers whose autonomy would be diminished. Within a few months, any possible lasting benefit from the effort came to naught.

One of the financial managers involved in the voting noted that the work performed normally would have cost about $150,000 and taken more than a year for completion. As a one-man endeavor the internal six-month effort had cost less than 20 percent of that amount.

Personally, I did achieve my professional goal to formulate a new intellectual base for future work. This was the last major think-piece that I did for IBM. All important papers and studies during the next decade were performed gratis for the Air Force, DOD, Congress and the White House — while still an IBM employee.

The end of 1972 and all of 1973 were very difficult times for everyone. President Nixon's effort to finally end the Viet Nam War began with Operation Linebacker II, December 19–21, 1972. We dropped 5,000 tons of bombs on North Viet Nam. We lost eight B-52s, two F-111s and several A-6s and A-7s in this part of the campaign. The North Vietnamese were "bombed back to the table" and negotiated a cease fire and peace treaty five weeks later.

This was followed by Vice President Agnew's resignation, the Watergate scandal that eventually forced Nixon to resign, and another Arab–Israeli war.

By 1970, IBM had clearly foreseen the massive evolution of the computer industry into integrated circuits and software, but the replacement of Vince Learson by Frank Carey as the IBM president in September 1972 did little to improve the company's position. IBM's stock price fell from the mid-$400s to the mid-$200s during the next two years.

In October 1972, the US Justice Department initiated an effort to break up IBM. This was followed by the *Telex* v. *IBM* lawsuit, in which Judge Christensen decided on a huge damage award to the Telex Corporation. The legal fees in these matters cost the Company up to $100 million a year for many years and pre-occupied IBM's senior management.

In late January 1973, FSD's evolution from the bomber avionics business to more emphasis on submarine programs warranted establishing a new facility for that purpose in Manassas, VA. Jackson moved Joe Jennette from FSD Marketing to the Gaithersburg facility to develop new civil business programs. Joe asked me to join him, but I had little interest in non-military work, I wanted to leave FSD Marketing entirely, and preferably I wanted to leave IBM altogether.

The only other IBM alternative was a move to the FSD Plans Department where Abe Katz was the director. A transfer was arranged. My first task for Abe was to analyze ESC-Owego's current plan. I demonstrated that their various program revenue estimates were inflated by 50 to 100 percent and the document was meaningless. Abe used the analysis mostly to gain perspective and never formally questioned the ESC-Owego estimated revenues.

At the meeting of an aerospace association in mid-April at Huntsville, AL (my first visit since the successful SECRAC proposal), I asked John Foster (formerly B-1 Avionics) from Hughes Aircraft about HAC's tiger list. He ruefully observed that HAC had thought they had the SECRAC win sewed up. They did not find out that I had rewritten IBM's SECRAC proposal until after the contract award.

Abe Katz transferred to IBM Corporate Headquarters, motivating Jackson to eliminate the rather ineffectual Plans Department. I was re-assigned to Earl Wells, now the FSD Controller. We both believed this to be a temporary arrangement as I was taking several initiatives to find a job in the Federal government.

Late in 1972, I advised Bob Evans of my efforts to leave IBM since I was unlikely to progress under Jackson and Bitonti. Evans tried to help me, principally through Gene Fubini, but Gene did not submit my dossier to anyone in DOD, only to the newly-formed Arms Control and Disarmament Agency (ACDA), an independent adjunct of the US State Department.

On April 30, John Walsh, a senior manager in DDR&E, said that he had nothing offer since DDR&E was facing a 14 percent reduction in force (RIF). He added that the current troubles of the Nixon Administration — Agnew, Watergate and now the Ellsberg/RAND leak of the Pentagon Papers to the *New York Times* — were contributing factors. He encouraged me to remain with IBM until things improved.

Many others put my name in for White House staff positions, the Department of Defense, and the secretary of the army for a position as assistant secretary of the army for research and development. The Defense Intelligence Agency, which had 300 people in the Pentagon and 1,200 more in other agencies and locations, was looking for a civilian counterpart to the military commander. In August, I became one of five finalists in that competition, but two of the applicants were colonels currently in the DOD intelligence structure.

In this period IBM and FSD senior managements were very privately offering special severance pay to selected older employees. The buy-outs usually consisted of a year or more of the employee's current salary, a few months pay while looking for a new job, and the IBM defined benefits plan when they reached normal retirement age. Joe Jennette had tried before, and on August 7, Jackson made an offer, but the terms were not favorable.

All my initiatives to leave IBM gradually failed. I was now a hostage of financial obligations related to my qualified stock options. It would have been preferable if IBM had instead given me one or two broken arms. My professional life returned to pedestrian assignments for Earl Wells. Since the Headquarters staff was substantially reduced, there were plenty of errands to do.

During the first trip to ASD-Dayton since the B-1 Avionics loss, I had a long talk with Jack Trenholm, including condolences for Paul Hockman's death in late September. Jack said that the B-1 Program was safe for now, but the later production authorization would be a very hard sell. He added that he wanted help with B-1 advocacy in the following year and would soon find some means for such an arrangement.

1973 Israeli War

On October 4, 1973, the Arab nations surrounding Israel launched a war to eliminate the Jewish state. With 2,000 tanks versus the Israeli's 700, and much larger armies, they almost succeeded. The Soviet Union had armed the Arab nations with those tanks and the SA-2 and SA-6 surface-to-air missiles (SAMs). The latter weapons greatly diminished the Israelis' air superiority and caused unsustainable losses of fighter-bomber aircraft. They lost 54 aircraft in just a few days.

The Israelis were holding on but the war was rapidly exhausting their munitions. A United States airlift re-supplied Israel and they continued their advance into the eastern Egyptian desert, the West Bank, Gaza and the Golan Heights of Syria.

The USSR was preparing to deploy 50,000 airborne troops into western Egypt. The US put SAC on a higher alert status and told the Soviets not to interfere. The US ordered the Israelis to halt their advances into Arab countries. The subsequent cease-fire lowered tensions between the US and USSR and ensured another 35 years of hatred, lesser conflicts, and usually futile negotiations with the Arab nations.

The Nike Hercules batteries had been removed from US cities during the mid-1960s due to the futility of air defense in the new era of ICBMs and SLBMs. These stored ASM systems, I thought, could be re-deployed to help the Israelis defend themselves, perhaps reducing the probability of a great-power conflict. This missile was still the most accurate in the world, for both surface-to-air and surface-to-surface attacks.

Israel had many post-war problems, but there were two that could be readily solved. The Soviet MIG-25 Foxbat Interceptor flew so high at Mach 3.0 that these aircraft could overfly Israel without being engaged by Israeli aircraft. Israel was such a small nation that their interceptors could not reach high altitudes before the MIG-25s departed their airspace. The severe attrition of Israeli fighter-bombers in the Egyptian desert was largely due to SA-6 SAM batteries defended by rapid-fire anti-aircraft guns. These air defense installations had finally been overrun by Israeli tanks, a very unusual and desperate means to re-establish air superiority. In the post-war years, these very effective defenses could readily be renewed.

I formulated a detailed plan for a Nike air defense for the Israeli nation using six Nike Hercules batteries (the same number required to defend Cleveland, OH) that would eliminate the MIG-25 problem. These installations could also be used to eliminate enemy air defense enclaves 100 miles beyond Israeli borders by ground-to-ground attacks. However, despite various attempts, there was no way for me to present these ideas effectively in the State Department. However, Major General Sumner, J-5 in the Joint Chiefs of Staff promised to explore my suggested plan.

Chapter 10. Air Force Invitational Orders

On December 3, 1973, Jack Trenholm called from the B-1 SPO regarding the Air Force's upcoming move to put the successful B-1 Bomber Development and Flight Test Program into production.

Jack wanted me to provide the assistance he needed through Invitational Orders, an arrangement that did not include compensation. The Air Force provided a very broad "need-to-know" on classified information, travel expenses, and unlimited travel on Air Force flights. Generally, the tasks were mutually agreed upon, but without any specific direction to the consultant. Such consultants usually were acknowledged experts in the field of investigation. The arrangement was very similar to the dollar-a-year man serving the government in World War II — but they no longer paid the dollar.

The USAF Aeronautical Systems Division chief of staff approved invitational orders (I/O) for me. Agreement was obtained from John Jackson and the FSD Legal Department to participate whenever the B-1 advocacy task did not interfere with obligations to IBM; occasionally, I would have two simultaneous full-time jobs.

A study plan was outlined for a visit to the B-1 SPO on December 13–14. Trenholm and the B-1 SPO Staff were pleased to have me onboard. Jack and Jim Sunkes reviewed and improved my study outline. Major General Douglas Nelson, still director of the B-1 Program SPO, also favored this new initiative. Jack established a liaison arrangement at AFSC Headquarters with Lt. Colonel John Thulin, reducing my travel time. Knowing the USAF Headquarters staff's resistance and hostility to any outside initiative by the B-1 SPO, Jack advised me to avoid the Air Staff.

B–1 Advocacy Study

During the next two months I structured the new study as a very broad examination of the total US strategic force structure, emphasizing the unique and necessary contributions of a B-1 Bomber Force in the overall Strategic Triad. I reviewed the justifications for the B-1 during 1962–1972, and their evolution to current Air Force positions.

Next I enumerated the vulnerabilities in the present AF advocacy. (1) The need for the Triad was either fiction or doctrinaire; there were no current prospects of a significant ABM defense or breakthroughs in anti-submarine warfare versus our SLBM fleet. (2) The unique and necessary mission was hard to sell, absent a clear counterforce policy, but this was the principal justification for the bomber penetration mission. (3) The USSR was placing much greater emphasis on strategic ballistic missiles than aircraft, and (4) The B-1 was competing with B-52 and FB-111 improvement programs.

The principal new factors influencing either the need for, or the cancellation of, the B-1 included: (1) B-52 operational experience and losses in Viet Nam; (2) evolution of the USSR and PRC target structures in numbers, hardness, mobility and their range distributions from the SAC bases; (3) SALT I arms limitations; (4) overall capabilities of US ICBMs, SLBMs and forward-based systems forces; (5) alternative BDA capabilities such as SR-71s with satellite communications and ballistic missile support; (6) co-existent US bomber and tanker forces and their basing structure; (7) projected B-1 operational status including numbers, deployment, supporting aerial tankers, and the assortment of probable B-1 weapons; (8) the fact that post-World War II foreign policies were still evolving, and (9) US foreign involvements were in a regressive cycle since 1968 and that was likely continuing into the mid-1980s.

The outline also suggested eleven mission phases and eight potential new roles and missions; eight things the Air force could do to help its own case for the B-1; and ten arguments for quantitative development to justify a significant B-1 Production Program.

The proposed study was better suited for long-term performance by a major "think tank" organization. Instead, the intention was for one man to produce significant results in six months while working part time.

I gave Trenholm two preliminary reports and made my next I/O visit to the B-1 SPO February 7–8, 1974. Heading for the taxi in the early morning darkness, I slipped on the black ice that had formed during the night and slid 50 feet down our steep driveway. There was no time to change clothes, so I arrived rather shaken, disheveled, and torn here and there. A shoulder injury had to be surgically corrected a decade later.

Trenholm introduced me to the new B-1 SPO chief, Major General Martin. We had the usual lunch in the ASD Executive Dining Room with Brigadier General Sylvester, the deputy for systems at ASD. Most of my time was spent reading classified documents for excerpts useful in the new study.

Jack arranged for regular contact with Jim Sunkes and Gus Augustine at the SPO for additional information. He also asked me to visit Rockwell International, the B-1 prime contractor in Los Angeles, for talks with Tom Nelson, their manager for B-1 operations research and suggested that the ANSER Corporation, now run by Jack Englund, would be helpful with data on the Soviet strategic target structure.

My visit with Englund on February 27 was very helpful; while there, I also met Colonel Bryant and Lt. Colonel Mudzo from the Air Staff Study Group under Brigadier General Lukeman. Our working exchange was very good, but the longer-term implications were not. Mudzo asked for a three-hour meeting in the Pentagon on March 1. My work for Colonel Emil Block in 1969 was still in their files.

The Air Staff people wanted me to put early emphasis on the B-1 Air-Launched Cruise Missile (ALCM) to support their current work on stand-off missile launching. Since I would eventually have to study the strategic ALCM role, I agreed. I produced a B-1 ALCM paper in a few days. As usual, I provided an unsigned copy to be used without attribution. Two weeks later, Colonel Bryant came over to my Rockville office to discuss the ALCM paper. He thoroughly liked the work and asked me to sign his copy. Jack Englund and Lt. Colonel Thulin at AFSC were also pleased with this exchange.

Jack Trenholm extended the I/O–1 Study Phase beyond the 1974 fiscal year and arranged for me to visit SAC Headquarters in Omaha, accompanied by Jim Sunkes from the SPO. He had received my ALCM study for the Air Staff, but since it was unsigned, he had not known the source.

I continued working on the counterforce exchange model and the equivalent megatonage (EMT) and counter-military potential (CMP) of all US and USSR strategic nuclear weapons. EMT calculations included the expected weapon delivery accuracy. CMP calculations were similar, but useful primarily for the high yield or very accurate weapons needed to destroy hardened military targets.

On April 17, an urgent call came from Trenholm. There was a "big rumble" with the Air Staff and SAC regarding my current work under the B-1 SPO invitational orders. Jack cancelled the SAC trip and asked me to stick with lower profile work, with only Lt. Colonel John Thulin at AFSC Headquarters and Jack Englund at ANSER until this matter was resolved.

Jack also said to avoid Colonel Bryant and other Air Staffers and to retrieve the ALCM notes from Lt. Colonel Mudzo. Thulin confirmed that this was a power play by the Air Staff, with me as Jack's consultant in the middle. I was concerned that there would also be negative feedback into IBM that would jeopardize the I/O arrangement. Jack Englund said that Colonel Bryant and/or Colonel Sperry were the source of our problem.

Thulin later confirmed the foregoing and noted that their boss, Brigadier General Lukeman, was also a factor. He added that things were going badly for the B-1 on the Hill, and offered to help me look for another job outside of IBM. I appreciated his concern during this very depressing turn of events.

B-1 advocacy was finally the responsibility of the Systems and Analysis Group in the Air Staff. Generals Lukeman and Welch were apparently more concerned about their prerogatives than results. This was a severe change in motif. Earlier, Colonel D.C. Jones of the Air Staff had been open-minded enough to carefully consider the paper I wrote for him in 1960 on the B-70 and eventually, but reluctantly, concurred with the conclusions. Similarly, the B-1 paper was very useful to Lt. Colonel Emil Block in 1969.

Studies

I continued working 100-hour weeks to complete the B-1 advocacy final report before the June 30, 1974 deadline. Major General Nelson decided the report contained so much sensitive, classified intelligence data that it would have to remain Secret for the next 30 years, until 2004.

Jack Trenholm called on July 17 to advise that my report was under review by Major General Martin and two others. The document would then be provided to ASD Systems Development, SAC and USAF Headquarters for their evaluation. I prepared the report briefing of 75 charts and spent several hours with Trenholm on August 21. General Martin had not yet approved the release of report.

Jack Cannon called from the ASD Plans Office on August 29, asking me to help evaluate "a B-1 paper by J.M. Collins of the Congressional Research Service of the

Library of Congress." The B-1 SPO had unofficially provided Cannon with my report. Trenholm similarly and unofficially sent the report to SAC and Lt. Colonel Thulin.

I went to ASD Plans (YH/XR) and reviewed the Collins paper for two days and nights. It was both poorly done and irresponsible. Jack Englund later found the Library of Congress report to be even worse.

Trenholm asked for a formal two-hour briefing of my advocacy paper on September 12. The audience included four colonels and ten civilians. The next day I repeated the story for Colonel Fellini and 30 others. These were the first presentations that I really enjoyed during the past four years. We later had a private session with Colonel Boice that also went well.

Jack called a few days later and asked for more copies. He said that SAC Headquarters had yet to be convinced. SAC sent their review summary to the SPO on October 25. They did not like the report.

We did not give up easily. Jack intended to set up a SAC briefing with some of the SPO staff early in 1975. He finally sent the advocacy report to General Welch in USAF Systems & Analysis and to General Darnstandler in DOD RDT&E. Thulin said that they thought it was great in DDR&E, but much less so in Welch's shop — as expected.

I visited the B-1 SPO on January 7, 1975, to see Jack Trenholm, Jim Sunkes and General Martin in preparation for the SAC briefing the next day. The first screening briefing to eight SAC field grade officers was like preaching to the choir. Colonel Hegenberger asked me to stay another day for briefings to Generals Beckhardt and O'Malley in DCS/Plans. The briefings went well and Colonel Kelly provided some good feedback. Glen Hadsell, a senior civilian at SAC whom I had known for years, asked for a return visit to SAC under his auspices.

SAC provided more authoritative data on the expected operational accuracies of all strategic weapon deliveries. I spent the next three weeks improving and refining the calculations and graphs. The story was now quantitatively better, but there were no significant changes in my original findings and conclusions.

On January 28, the B-1 SPO chief, Major General Martin finally agreed to hear a two-hour briefing of the advocacy report. Since I had submitted the document seven months earlier, this was an indication of his wariness of the findings.

On February 5, Jack told me to distribute the report as a personal initiative, avoiding any contact with Brigadier General Welch. Field grade officers were contacted in the various commands and the civilians in DOD RDT&E. The latter office advised that the B-1 Program was absorbing 17 percent of the Air Force budget for all such developments.

I received many calls indicating that the report was widely read and appreciated at the working level. Further, parts of the document were included in statements to the Congress by the Secretary of the Air Force and the Air Force Chief of Staff. This made the 14 months of overwork seem worthwhile. We were still a long way from B-1 production, but the report was a solid basis for the next attempt.

The B-1 invitational orders provided me a much-needed sense of purpose in my professional life, but I still had to work for Earl Wells. This evolved into a three-year pattern of writing the top-level FSD planning document, a 100-page report, during April and May of each year.

Earl also asked me to write the executive summary for the FSD Five-Year Plan; this five-year plan summary required three months in the fall of each year.

I also consulted on major FSD programs, such as the World-Wide Military Command and Control System-II (WWMCCS-II), B-52, B-1 and anti-ballistic missile defense studies. All this added up to more than a full-time job.

World-Wide Military Command & Control System-II (WWMCCS-II)

Even the acronym for the World-Wide Military Command & Control System-II had to be shortened to WMX for working purposes. The initiative was "-II" because the first study had been ineffective. The program was a very ambitious high-level Pentagon effort to consolidate all warning, communications, intelligence, data processing and battle-management systems into a centralized, world-wide command and control system for the military establishment.

DOD took the macro, top-down approach to selecting a prime contractor for the study. Historically, because of its great size and commercial reputation, IBM did well in such initial competitions. John Jackson, Vince Cook and others made a Pentagon presentation that promised DOD the best of everything in FSD and potentially great support from IBM Commercial. They won.

Vince Cook recruited people from throughout FSD. Nearly all of the ESC-Owego Operations Research people were eventually moved to a newly leased facility in Rosslyn, VA. Cook also insisted, for several months, on my transfer to his program. Jackson wanted me to participate as well; I offered consultation services. First, Cook wanted me to review their first major report to DOD on the WMX Pilot Architecture. The document was useless and I advised Cook not to release it to DDR&E.

Because of my professional background and current work for the B-1 SPO, I knew many things that could be useful to WMX. Although I refused to brief Cook or McCrae, I fed such items to their study at the working level, usually through Steve Wilson, who had become the WMX Operations Research Manager.

Vince McCrae was soon replaced by Dr. Al Babbitt, a highly regarded senior professional, with whom I always enjoyed working. The IBM WMX Study continued for about two years, eventually declining to 20 people before termination. DOD has made great progress toward the WMX goal during the subsequent 30 years, but this was a generational problem that could not be solved by top-down studies and quick fixes.

Renewed Orders

Jack Trenholm renewed my invitational orders twice during 1975–1976 to ensure continuing contacts with Air Force officers concerned with B-1 advocacy. The first phase of the Orders (referred to as I/O–1), was a useful contribution to B-1 advocacy and my efforts were well received, but not enough to ensure a B-1 Production Program.

Jack acknowledged my planning document obligations to IBM and did not expect a major effort until September, with a new and final report in early 1976. My first effort had provided a useful appreciation of Soviet strategic weapons growth to eventual superiority in megatons, equivalent megatons, and counter-military potential (MT, EMT & CMP; respectively). However, there was not enough time or data to calculate the more useful timing and significance of their strategic armaments programs.

The summer months allowed collection of the additional data needed on US and USSR strategic missile and aircraft forces with their opposing target structures. ASD,

AFSC, SAC and the ANSER Corporation (Jack Englund) were all very helpful. They were motivated by the progress I had made during my I/O–1 effort. Englund also advised me that IBM was now considered a third-tier subcontractor in the strategic bomber business due to our fiascos in the B-1 and B-52/SRAM competitions.

US & USSR Strategic Forces Study

The major quantitative advancement from my I/O–1 study would be to determine what the two great powers could do with their evolving strategic forces and when these capabilities would be realized. While the British and French nuclear forces were a rational deterrent to any attack on their countries, they were relatively insignificant by themselves. The Soviet strategic forces eventually became so powerful that if they were committed against either country, they could turn everything above ground to cinders in less than an hour.

Being a nuclear superpower required both large, diverse strategic forces and a very large land-mass. Even so, in a general nuclear war both the US and USSR would experience an estimated 150,000,000 casualties. (At that time, the United States had about 220 million and the Soviet Union encompassed roughly 270 million.) This single fact created a common deterrence policy by default — Mutual Assured Destruction. This very unpopular and dangerous policy precluded nuclear war between the superpowers for half a century and continues today. The current status of US-Russian nuclear parity is defined by a brief unclassified study that I performed in 2008 for the final chapter of this book.

Counterforce Model

To offer anything new and significant, my I/O-2 Study required that I construct a mathematical model for the all-out commitment of US and Soviet strategic missiles and bombers against all of the opposing strategic target structures. The industrial/urban (I/U) target systems had been established for 25 years; they simply grew larger in area and slightly more numerous. Such targets were not time-urgent and could be readily destroyed by any nuclear-armed weapon system. Consequently, the least capable weapon systems would generally be assigned to these I/U targets.

The critical initial targets for a counterforce exchange were the strategic airbases, ICBM silos, associated launch control centers (LCCs), command and control centers, and nuclear weapon storage facilities. In later years, the hardness of these targets required 3,000 PSI or greater overpressures from the nuclear weapon(s) to ensure their destruction. This warranted emphasis on the higher weapon yields and greater delivery accuracy defined by counter-military potential (CMP).

Submarine bases were also high priority targets to ensure destruction of the SLBM boats before they could get to sea. If already at sea, these submarines could launch their missiles against time-urgent, semi-hardened targets in less than the half of the flight time of an ICBM. The nominal 12- to 15-minute trajectory of SLBMs made them vital weapons for attacking airbases and the command and control centers.

The availability, reliability, survivability and progressive deployment for each of the opposing weapon systems had to be consistent with the best available intelligence estimates. Following the US completion of 1,000 Minuteman ICBM deployments and the retirement of our older Atlas and Titan ICBMs, American land-based ICBM forces were essentially static in numbers. There were continuing improvements in accuracy and the evolution of the multiple re-entry vehicles (MRVs) to more versatile maneuvering independent re-entry vehicles (MIRVs).

United States ballistic missiles were always more accurate and reliable than the comparable Soviet ICBMs and SLBMs. This was largely due to the greater precision of the gyroscopes and accelerometers in our inertial guidance systems. The Soviets more than compensated for our technical superiority by deploying greater numbers of missiles with much larger nuclear warheads. American SAC bomber forces were always more numerous and effective than those of the Soviet Long Range Air Arm (LRAA).

The US and USSR had developed very sophisticated orbiting satellite systems with infrared sensors to detect ICBM and SLBM launches. The satellite sensing systems evolved to capabilities that could analyze the initial trajectories to predict the target locations and impact times for each and all of the attacking missiles.

The infrared satellites were backed up by Ballistic Missile Early Warning Systems (BMEWS) for the US and comparable ballistic missile radar detection systems in the Soviet Union. Our long-range radar installations were in Greenland, Alaska and Great Britain. The Soviets had similar facilities on the northern periphery of the USSR.

The aforementioned satellites and radars provided certain warning of the launch of all ballistic missiles. If the Soviets attacked first, these warning systems would give us precious time to launch our strategic forces before they were destroyed by incoming Soviet warheads. These circumstances led to a US counterforce doctrine known as SLUCA, Strategic Launch Under Confirmed Attack. This required and allowed the immediate launch of US strategic bomber and missile forces.

The near simultaneity of the strategic counterforce exchange provided the fundamental basis for my counterforce exchange model. In the late 1970s, this required selectivity in both the US and Soviet attack sequence between time-urgent military targets, such as ICBM silos, and other strategic targets that included I/U complexes. As the Soviet forces grew in numbers and capabilities, they would be able to attack all strategic targets simultaneously by the mid-1980s.

With our smaller strategic forces and the immense Soviet target structure of several thousand locations, the US did not possess this capability. Building the counterforce model required that I calculate the probability of kill for each type of nuclear weapon for each type of hardened target. This included the reliability and availability of each weapon system, progressing over 15 years from 1970 through 1985.

The opposing target structures were also grouped according to importance and time-urgency. Not having access to a large computer and supporting programmers, I used a hand calculator.

The opposing SAC and LRAA bomber forces were assigned to non-time-urgent targets, both hardened (nuclear weapon storage sites) and relatively softer targets such as I/U complexes. Our bombers were also to overfly all previously attacked hard targets to do BDA and re-strike missions. After 18 years, this was still the principal unique and necessary mission of the manned strategic bomber vis-à-vis our strategic ballistic missiles.

Ensuring the destruction of all silos and residual missiles that failed to launch or those missiles in storage were vital missions for the longer term prosecution of a general war. The Soviets generally had several hundred ICBMs in storage to reload any intact launch facilities. The US similarly stored over 100 Minuteman Missiles for that purpose.

The public dialogue on general nuclear war very seldom went beyond the hypothetical initial exchange. But after more than 300,000,000 people were killed, in the worst event in human history, neither side would be likely to simply declare victory and sign a peace accord. If they were not destroyed, ICBM and SLBM residuals and spares augmented by renewed satellites or other reconnaissance could readily terminate any substantial effort to re-arm or rebuild. The same capability would exist with any remaining bombers in SAC or the LRAA.

Next I set out to calculate a complete counterforce exchange between the superpowers at the end of each year from 1975 through 1985. The yearly increases in strategic offensive capabilities on both sides required separate and more complex assessments for the subsequent decade. The last year considered was 1985, because the Soviets would by then have sufficient forces to destroy 90 percent of all the military targets and all of the 150 largest cities in the United States. There was nothing more to be proven by further calculations. This circumstance would obtain for the foreseeable future.

My calculations for 1975 indicated that the Soviets had already surpassed the US in total MTs, EMTs and CMP. They would certainly maintain this posture, and would achieve the overwhelming counterforce capability in 1985. The Soviets were clearly seeking military superiority, while the US posture and policies were structured for the lesser forces needed in assured destruction for deterrence, and if necessary, devastating retaliation. The Soviets were in a strategic arms race and we were not! These analytical results would surprise most of the military establishment and would have been very disturbing to US citizens had the final report ever been made public.

Jack Trenholm extended the I/O-2 Study schedule into 1976, allowing me more time for the extensive calculations and the final report. Since I was working alone, any mistake in calculations or judgment in an earlier year would be reflected in subsequent years. To avoid that, I developed a technique for progressively plotting illustrations on all aspects of the study to detect any anomalies in developing trends. These many charts allowed me to write the final report in just two weeks, after months of calculations.

The final draft of the report was completed and taken to the B-1 SPO on January 12, 1976 for classification. They were concerned that so much Secret intelligence data had been integrated and such significant conclusions drawn that the report might be Top Secret. The SPO deferred to the ASD Foreign Technology Division. FTD decided that the report could remain Secret, but could not be declassified for 25 years. That decision made the document much more useful for a B-1 advocacy campaign by ensuring a wider military and government civilian audience.

The final report, entitled "Air Force Study of US and USSR Strategic Forces, 1975–1985," was a 178-page document with 38 illustrations. My immediate audiences opined that nothing like this document had been previously written, let alone by one private citizen in less than eight months (including the earlier preparatory effort in 1974).

On January 15, 1976, a 60-pound package was delivered to the B-1 SPO. I had to book an adjoining seat on the plane for the classified reports. With this work in Jack Trenholm's capable hands, I enjoyed the ride home — while waiting for the pot to boil.

The new report caused a stir in the US intelligence community. The results proved what many professionals already suspected. Further, the study was thorough and conservative, with much care taken to avoid creating new intelligence data. I described my methodologies, to allow the Community to verify the results. Major new insights had been provided into the significance of the rapid expansion of Soviet strategic forces and what the United States could and must do to appreciably offset their growing superiority.

I recommended three major initiatives to partially offset the Soviet buildup: (1) continuing emphasis on great vigilance to ensure the Soviet perception that the US would launch a devastating retaliatory response to any Soviet surprise attack, particularly in the fall of each year; (2) placing the B-1 Bomber into full production as the only new program currently available to partially offset Soviet superiority in MT, EMT and CMP (no surprise there); (3) accelerating development to allow earlier production of the new MX ICBM and improve our capability to rapidly destroy the ever-expanding Soviet hardened target structure.

After 24 years of experience in advocating the results of unpopular positions and studies on Nike, B-70, B-52, SAGE, and the B-1, I knew the final report had to be widely distributed and briefed before the findings would cause an appropriate government response.

There were professionals in the intelligence community who could have done this study eventually, but they would not have been able to make themselves heard through the bureaucracy. Consequently, this initiative as a private citizen was unique.

Jack Trenholm rapidly and selectively distributed the report to AFSC Headquarters and to Jack Englund at ANSER. Jack called to say the B-1 SPO had asked him to evaluate the study and to verity the results. He told me that the only way he could do that would be to again use Top Secret data.

Trenholm asked me to prepare briefing charts for the entire study during February. A minimal trial briefing was given to Earl Wells on February 27. He was quite taken aback by the scope and impact of the study. On March 2, the first formal (screening) briefing was given to Trenholm at the B-1 SPO. He was determined to take action on the findings.

Trenholm arranged a briefing to the B-1 SPO Chief, Major General Martin, on March 18. Only two months delay this time, versus seven months for I/O–1. I briefed the General and 40 other invited officers and civilians from the ASD Headquarters for two hours. General Martin then met privately with the two of us. He intended to endorse the story to AFSC, SAC and USAF Headquarters. He also arranged for briefings to Major Generals Larson (Systems Development) and Dunn (Plans) on March 31 at ASD Headquarters.

AFSC Headquarters asked for ten more copies of the report and scheduled the briefing there on March 31 to a brigadier general and seven colonels. All of the presentations through March were well-received.

Lt. Colonel Bernie Miles called to schedule the USAF HQ briefing to Systems and Analysis, now run by a non-admirer of mine, Major General Jasper Welch. I remembered Miles as a bright young Air Force Captain during the A-7D/E competition nine years earlier. We got on well.

Jake Welch was briefed as scheduled; he questioned or criticized every aspect of the report. Bernie called later to say that the Air Staff generally thought the report was both valid and useful. He added that General Welch's reaction to it was entirely "NIH," that is, "not invented here." Welch was a bright young officer providing direction to a large and important portion of the Air Staff. He was responsible for, among many other things, the advocacy and progress of both the B-1 Bomber and the MX

ICBM. He understandably resented a lone pilgrim providing a better story for both systems than he had established, with all of his resources and prerogatives.

During my next trip to the B-1 SPO, Jack Trenholm asked how IBM liked the letter that the ASD commander, Lt. General Stewart, had sent six months earlier to John Opel, the President of IBM, thanking IBM for my considerable services to his command. I could only reply, "What letter?" Jack offered me his copy, and Opel's one-sentence reply thanking General Stewart for his letter. This exchange showed Jack how little IBM cared anymore about strategic bombers.

The Watergate Affair resulted in the resignation of President Nixon in August 1974. President Ford replaced him and a few months later shook up his cabinet by replacing Secretary of Defense Schlesinger with Donald Rumsfeld. Dick Cheney became Ford's new Chief of Staff.

Several senators were very vocal in their views on bombers and the Democrats in general were very much opposed to any B-1 Production Program. However, professional sentiment in the military was shifting toward a stronger defense posture. My study provided a solid analytical basis to justify such changes. On April 22, a new and classified SLUCA Doctrine was established. This was entirely consistent with the counterforce model developed during my study, but I do not know if there was actually any connection between my study and the policy change.

The critical test of the report would be the SAC Headquarters briefing on the morning of May 5. Colonel Northrop had provided an audience of about 40 senior field grade officers and civilians for the one-hour briefing. Briefings could be cut in half at SAC; they collectively knew more about everything than the speaker. The screening briefing was successful and they endorsed it to the general officer level.

That evening I briefed Major General R. Cody and six of his colonels. The General, relaxing with his evening glass of wine, was both receptive and complimentary. He encouraged the propagation of the study in the Air Force and DOD. This was a big relief after the beating I took from General Welch at USAF HQ. I advised Trenholm and Bernie Miles of the favorable outcome. Trenholm also approved of my first recommended initiative in response to General Cody's approval.

There was to be a luncheon meeting on Capitol Hill, sponsored by the defense industry, where the Air Force Chief of Staff General David C. Jones was the speaker. General Welch was there too, but less than pleased to see me. After General Jones gave his speech, and 14 years since our last meeting, he nodded his recognition.

I stood aside while General Jones answered questions; then he came over and we shook hands. I let him know our meeting was by design and offered another report that I thought he would find interesting regarding the B-1 and MX ICBM.

General Gray called with thanks for my efforts and said that, after General Jones' review, he had sent the report to General Welch. Jake was not likely pleased to receive his second copy, but he would likely be more open-minded this time.

Major General George Keegan, Air Force assistant chief of staff for Intelligence for many years, was an effective defense activist. He selectively provided Top Secret, broadly based intelligence briefings to the senior managements of large companies such as IBM. His mission was to inform the Defense Industry of the evolving Soviet threat. By good fortune, I attended his three-hour briefing on June 15. At the Top Secret level, his briefing was both more authoritative and, in some ways, scarier than mine, but the two stories were entirely compatible.

Trenholm asked me to set up a meeting with General Keegan. He also arranged a briefing to Dr. Antonio Cassioppio, the venerated chief scientist for the foreign technology division at ASD, on June 23. "Tony" and six of his department heads first asked a long series of questions about the data sources and methodology. Then they settled into 90 minutes of silence for the briefing. Dr. Cacciopo said that General Keegan was the Air Force's brightest general and that he should hear this story. He called immediately to set up a meeting with General Keegan on July 15.

Following several days of intense preparations, I briefed General Keegan at the Pentagon. The small audience in a windowless, dark conference room included Colonels Sell, Morris and others. One major also attended, Verne V. Wattawa. Our paths crossed again eight years later and we became good friends for life.

Keegan asked that all of the briefing charts, reports and supporting calculations remain with him. He said that he would call me shortly.

Then I took Barbara on our first vacation in three years. While we were at Rehoboth Beach, Colonel Emil Block called to advise that General D.C. Jones had assigned him as the new honcho on B-1 advocacy. I filled him in on all of my recent advocacy activity. A meeting was also scheduled with Colonel Sell to talk with General Keegan on July 26.

The hour-long meeting with General Keegan had the best possible outcome. He said that his staff had verified all of the calculations and that the conclusions were valid at all levels, Top Secret and above. He said that the Secret data had been used so carefully that otherwise Top Secret conclusions had been deduced. George was in a generous mood; he added that "the approach was brilliant and the results beyond challenge."

General Keegan's good mood was likely due to the results that he obtained using the report. He had met with senior Air Force Generals and Dr. M. Currie of DOD DDR&E. Their new intentions were still classified, but due to this analysis, the Ford Administration intended to put the B-1 Bomber into production and to accelerate the MX ICBM Development Schedule by two years!

General Keegan also requested I serve as a voluntary consultant to his office for the rest of 1976. He then retired as the longest-serving general as the chief of Air Force intelligence.

Since I was only cleared for Top Secret access, his staff would bring me unmarked papers to review. When I was called in for a two- or three-day assignment, this work was done alone in an austere, windowless Pentagon office. My broad and detailed knowledge of the Soviet Rocket Forces and proven methodology provided the expertise necessary for certain analyses and reviews. I never discussed this subsequent activity and arrangement with anyone.

Concurrently, I also briefed and consulted with Colonel Emil Block as he moved the B-1 Bomber toward initial production. My last formal briefing to the Air Force under I/O-2 was on November 4, to Major General Patton, DCS for Intelligence (AFIN). Colonels Block and Katz and Major Wattawa attended this meeting.

Chapter 11. Program Management & Congressional Liaison

The final success under Air Force invitational orders came at the expense of tolerating very poor professional circumstances in IBM. Jim Bitonti moved to FSD HQ in May 1974 and Ken Driessen became general manager at ESC-Owego.

As I prepared the BS&O Report for 1976, I ran into the usual production problems. Earl Wells' people contributed only five pages and they were late, inconsistent, inaccurate and incomplete. This time, I had a surprisingly bitter exchange with Wells. He asserted I would no longer be writing the BS&O or the Executive Summary of the FSD Five-Year Plan; it would be transferred to a newly established planning function headed by McCrae (whom I had earlier identified as a problem, not an asset, on the WMX program). This effectively eliminated my job. This was the first deliberate move since 1973 by Jackson, and now through Wells, to effect my separation from the Company.

I made it clear that I would not voluntarily separate from IBM without an IBM Corporate review of the current circumstances and my considerable prior contributions.

On September 30, Earl told me that Jackson's new assignment for me was to "determine why the FSD Command and Space Center was coming apart." Former FSD president Bob Evans had built the CSC in Gaithersburg, MD, to establish a major facility closer to Washington, DC, and our federal government customers. Evans, and later Jackson, developed it into a software- and systems-oriented facility that sold IBM commercial products through many FSD facilities throughout the United States. FSD personnel from at least eight other locations were assigned to report to the CSC management; this was a very unwieldy structure.

During October 1976, I interviewed 23 senior, middle and marketing managers at CSC for a total of 35 sessions. A management study of this scope would ordinarily require several months. Likely more interested in my failure than in any constructive study results, Jackson demanded the final report in one month. However, they seem not to have read the report in the first six weeks after its completion.

The 41-page report provided them with many more insights than they expected or wanted. The principal conceptual recommendations included:

1. Consolidation of Civil Space management at Houston with supporting new business and performance elements at Gaithersburg, KSFC and Owego;

2. Near-term elimination of the Huntsville facility and distribution of these assets to other CSC and Owego locations;

3. Consolidation of CSC Gaithersburg to include Westlake and certain business area assets at Manassas and Owego;

4. Extensive restructuring within CSC-Gaithersburg to emphasize command, control and communication; civil and military space programs; aerospace surveillance; systems engineering and integration; and software.

The first response was from Joe Fox, CSC general manager, on December 1; he generally agreed with my recommendations. Wells was less pleased. Jerry Ebker initially reviewed the report for Jackson and Bitonti and asked me if I thought it would be best to remove Joe Fox as CSC general manager. I had assured all the interviewees that I would not offer any such opinions.

During 1977–1978, FSD senior management acted on several of my recommendations. Jackson eliminated the Huntsville facility and gave the buildings to the University of Alabama. The reporting channels were changed as I recommended for the KSFC and Westlake; CSC new business activities were realigned; a functional software organization was formed in Gaithersburg; and CSC later had a new general manager.

My three-year I/O obligation to the Air Force was now successfully completed. I was finally vested in IBM's defined retirement benefits plan. At last I had the latitude to take greater risks. No longer working for Jackson, Bitonti, Wells and IBM would now be a relatively small loss and a great personal relief. Still, I had to try to get a better position in IBM, however unlikely that might be. I submitted a letter requesting re-assignment in FSD on December 10, 1976, and noted I had endured many years of employment far beneath my proven potential in both workload and responsibilities.

The attachments included a summary of my major accomplishments in IBM. Jackson appreciated that this package could be just as readily sent to IBM Corporate Headquarters, should such a terminal move become necessary.

B–1 Computer

Jack Trenholm recognized that his 1975 effort through General Stewart to draw IBM President Opel's attention to the value of my Air Force contributions had been a non-event. This time, Jack tried another approach that could somehow benefit IBM and possibly me.

When Boeing won the B–1 Avionics Contract in 1972, they had a very inadequate computer subcontractor that gave the B–1 SPO many problems during B–1 development and flight tests. Consequently, the SPO staff had spent several months preparing a new preliminary B–1 Computer Specification to re-compete the anticipated production program. On February 2, 1976, Jack sent me the preliminary 100-page B–1 Computer Specification through an Air Force intermediary. Trenholm intended that IBM would use this information to eventually re-bid for the B–1 Computer, thus improving both the B–1 Avionics performance and my professional situation. My demotion was obvious; Trenholm was aware of it, but I never discussed my IBM problems with him. He did not appreciate how unlikely it was that I would ever again work on the B–1 under current FSD management.

The basic IBM B–1 computer requirements, specification and architecture had been developed when I was managing Department 566 under the ASSB Study Program in Owego during 1965–1970. By late 1966, the design was complete for the several such computers needed in each B-1 Aircraft.

When IBM bid for the Space Shuttle computers in the early 1970s, the B-1 design provided the basis for their proposal. FSD-Owego won the Shuttle computer contract and designated their new product the AP–101 Computer. Each Space Shuttle carries five of these computers and a "hot spare" that can be immediately substituted for any active computer that fails. Through very complex computer programming, the Shuttle can return to Earth safely if any two of the six computers are functional. If they were so inclined, ESC-Owego could readily have re-adapted the AP–101 Computer to the B-1 application, using more modern technology developed during the subsequent decade.

I gave the preliminary computer specification to John Jackson on February 3. He fully appreciated the immense market potential and unique opportunity. He immediately called Ken Driessen and sent the specification to Owego. ESC-Owego did not respond.

John sent me there on February 19. The visit stirred resentment at all levels of management, but it got their attention. I made a second visit to Owego and prepared an outline of what had to be done. They committed to make a presentation to the B-1 SPO on April 2. The meeting was a failure due to the Owego technical personnel.

Boeing finally distributed the same B-1 Computer Specification to the avionics industry on August 23, 1976. The insurmountable advantage that Owego could have realized with a $100,000 six-month pre-proposal effort had been thrown away.

Jackson required me to go back once more on October 15. Driessen refused to bid because the proposal and follow-up was now estimated to cost $535,000. Years later, FSD received a B–1 computer contract as a subcontractor to Boeing.

The B-1 SPO arranged a briefing of the I/O-2 Study to the senior management of the B-1 prime contractor, Rockwell International, in Los Angeles on June 24, 1976. Following the unpleasantness over the B-70 Program, we had always gotten on well. The audience included Warren Swanson and Tom Nelson. They had followed my activities in B-1 advocacy over the industry grapevine for years and were very supportive.

Many years of IBM employment, volunteer work for the Air Force, helping Barbara raise and educate our son and daughter and care for aging parents, financial pressures and other factors had caused me long periods of severe stress and frequent depression. The most frequent problems were sleeplessness, periodic random and severe chest pains, bruxism (tooth grinding to the point of breaking molars), and scintillating scaratoma — essentially optical migraine.

I had developed a technique to manage pain and stress during World War II in the Marines. At Camp Lejuene, NC, in 1946, a library book described a technique that was essentially self-hypnosis to effect post-hypnotic suggestion. A milder form I practiced in later years might be called meditation. A few minutes of self-hypnosis could have the same effect as several hours of sleep. One example was the last and all-time maximum 138-hour workweek in finishing the A-7D/E Proposal in 1966. Our deadline was less than 24 hours away and our proposal management team worn out. Late in the evening, I had John Cooney schedule a new secretary for every three hours all night long. After about 15 minutes of self-hypnosis, I dictated the Executive Summary of our proposal in about 10 hours.

Technology Transfer

January 1977 began with several meetings in which John Jackson and Earl Wells stated the same severance terms that they had tried to impose three years earlier.

I sent my resume through carefully selected intermediaries to the CIA, DIA and Air Force Intelligence; but the new Carter Administration, my high salary in industry, and 51 years of age (reality often transcends the law in such matters), were all negative factors.

Joseph L. Brown, formerly a very senior manager in Commercial IBM mainframe computer manufacturing, was now the FSD Headquarters director of technology. Joe came to understand that I preferred to be in Technology. Jackson transferred me and Joe agreed that I would work directly for him.

Joe Brown was my first really good technical manager in many years. He soon recognized my ability to do anything related to his job — a demotion that he thoroughly disliked. He told me to get up to speed on the new semiconductors that had developed during my seven-year exile in marketing and planning, and assigned me many peripheral tasks, including running FSD Technology during his frequent travels, attending all FSD Headquarters DCAs, critiquing all new FSD plans, answering much of his mail, participating in the Interdivisional Technology meetings and all relevant national industrial organization meetings. I also arranged a seminar for 200 FSD engineers and managers at UCLA and conducted a year-long FSD-wide Display Strategy Study that we finally briefed to IBM-Armonk.

FSD continued to evolve around us. Likable Joe Fox resigned as general manager of CSC-Gaithersburg on May 5, 1977. He was replaced by Vince Cook. As Joe Jennette remarked on his way out, "Some days you just have to leave early."

I advised Don Spaulding and Bob Evans of my recent Air Force work. Evans wanted a briefing on the I/O-2 report concurrently with Gene Fubini. Evans was impressed with the scope and findings of the study, but more so by Dr. Fubini's complete agreement with the conclusions. Gene immediately set up a briefing to Andrew W. Marshall, for many years the Director of Net Assessment for the Department of Defense.

Evans later asked for a review of my last decade with FSD. I noted that the three-year I/O effort had culminated in the B-1 Production Program and two-year acceleration of the MX ICBM Development, but that I had not been given the required IBM performance appraisal in seven years — not since the disingenuous review by Bitonti and Stoughton in 1970.

As an aside, Bob noted that Joe Fox, then general manager of CSC, had been a protégé of IBM President John Opel. Opel's replacement by Frank Carey made Fox's tenure less certain. Progress in IBM upper management, as in most organizations, required a senior sponsor, casually known as a "Daddy Rabbit." Bob Evans was the closest to that role in my case, but his consistency left a lot to be desired. Evans told me to defer any job changes for 30 days.

The briefing to Andy Marshall and his staff went well and he wrote a summary to Fubini on March 28. Andy thought that many of the trends that cited were valid, but the one-man operation was too limited to address the "uncertainty bands in estimated yield, accuracy, reliability and operational uncertainties." He noted that my annual dynamic analysis of counterforce capabilities was very innovative and valuable and added that the innovation should be adopted in other larger and broader studies.

On April 7, Earl Wells, still my manager, pending transfer to FSD Technology, presented me with an FSD Outstanding Achievement Award of $5,000 and a six percent salary increase.

MI 10-24

When Jim Bitonti transferred to FSD Headquarters, we tended to avoid each other. On February 28, Joe Brown said that Bitonti wanted him to define an FSD internal audit function to identify problem areas in newly awarded programs and advised me that a new FSD Management Instruction, MI 10-24, was being formulated. With the declining FSD population and various FSD management problems, there were deficiencies in many of our programs.

Joe insisted on my immediate involvement in both writing and implementing it. He said, "You are to be our conscience in FSD." We certainly needed one. MI 10-24 required that all contractual proposals or internal development programs would be supported by a Program Management Plan (PMP) for the duration of the program. The document was to be prepared prior to the business area Defense Contract Authorization (DCA) meeting. The PMP was continuously revised to reflect progress or changes during the performance interval.

A Program Control Review (PCR), essentially a broad internal audit, was to be conducted within 60 days after program initiation for all major development programs, with a PCR team to consist of FSD HQ staffers from plans, contracts, marketing, operations, pricing and selected technical consultants. The PCR would verify the validity of technical plans; adequacy of funding, staffing and controls; and review the PMP. Early and prompt changes in program managers or technical staff following a poor PCR soon reinforced the process. I conducted 77 such PCRs during the next six years. My new job became the most influential non-management position in FSD. When I insisted on re-assignment in 1982, the PCR process rapidly disintegrated.

Defense Science Board

As a result of the successful briefings to Evans, Fubini and Andy Marshall, I accompanied Bob to the Defense Science Board (DSB) meeting in San Diego, during July 10–17. The DSB was holding a joint meeting with the Air Force Scientific Advisory Board to study the cruise missile. Air, sea and ground-launched variants would all be studied along with their expected technology advancements.

Evans wanted to ensure our greater contact with senior people in the defense sciences, propagate the views established by the I/O-2 Study, and benefit from the many high-level presentations. His other agenda item was that I could cover for him while he missed many of the meetings due to the Government's legal action to break up IBM. There was a separate suite full of IBM lawyers just down the hall.

The DSB Cruise Missile Summer Study began with DARPA briefings on a new engine, autonomous terminal homing, and early stealth technology. The attendant DIA and CIA presentations on current and projected Soviet low altitude air defense capabilities were already well known. The Navy briefed on the sea-launched cruise missile (SLCM) and the Air Force covered the air- and ground-launched cruise missiles (ALCM & GLCM) and expected variants. SAC presented the operational concepts for all cruise missiles.

Evans was responsible for drafting a major portion of the DSB final report. This was the last part of his agenda. I spent the rest of July on that task and sent the report to him at IBM-Armonk.

B–1 Program in Congress

Soon after his election, President Carter and his willing helper, the new Secretary of Defense, Harold Brown, began to dismantle the B-1 Bomber Production Program. This negated my 15 years of advocacy and the previously successful effort to ensure B-1 production. Their simplistic rationale was that the B-52, performing stand-off ALCM launchings, would fulfill US needs for a strategic bomber force.

General D.C. Jones, as Air Force Chief of Staff, initially worked out a compromise on January 28 that would reduce the initial production from 15 to 8 B-1s. On September 5, 1977, Harold Brown announced that he and President Carter had terminated the entire current production option on the B-1, contrary to the recommendations of their Joint Chiefs of Staff. The B-1 Program would be limited to the five developmental aircraft now near completion.

Jack Trenholm encouraged my involvement in anything in Washington that would help the B-1 Development Program and eventually re-start B-1 production. We talked frequently about efforts to preserve at least the sixth prototype — the production prototype that would incorporate all the changes suggested by test experiences with the five developmental prototypes — as a means to eventually get the B-1 back into production without losing all that had been invested.

The B-1 Program had taken Trenholm well beyond his preferred retirement age. He advised me on August 16 that he would be retiring from the B-1 SPO because he could not professionally outlast the B-1 hiatus during four or more years of the Carter Administration. Since there was no longer anything to be accomplished by B-1 SPO visits, we conferred from our homes during the Carter years.

DOD initiated a new Bomber Study on February 3. Drs. Selin and York were participants. Bill Perry had been appointed to run DDR&E under Harold Brown. Admiral Stansfield Turner had been appointed to head the CIA, leading to the retirement of several hundred experienced agents.

Trenholm, then still with the B-1 SPO, asked for a review of the DOD Bomber Study Final Report. Colonel Barry provided a copy at AFSC Headquarters on May 21. Much of my time was spent on the report through May. At 4:00 a.m. on June 1, the review was completed. The new critique and analysis could refute the Bomber Study rationale on all critical points of their arguments. However, we were playing with a stacked deck against a doctrinaire pacifist administration.

On June 2, I took the Bomber Study Critique to Colonel Barry and he sent it to the B-1 SPO on the secure data network. Trenholm soon asked me to distribute the report to any organization in Washington where it might be helpful, including the Congress.

Bob Olds was a highly respected retired Air Force general officer then on the staff of the Senate Armed Services Committee. We had another useful two-hour discussion on the findings in early June. He advised that if the House of Representatives voted on it, they would likely kill the B-1 Production Program by 202 to 199.

The House leadership intended to avoid a full House vote and let the B-1 Development Program be reduced to six test and prototype aircraft. President Carter's action, mentioned earlier, limited the B-1 Program to five aircraft. Had we been able to save the sixth prototype B-1, eventual restart of production could have been accomplished in much less time and save about $1 billion.

Olds and other Air Force officers agreed that I should write an unclassified paper to be used to argue against the defense policies of President Carter. "The Impending Strategic Military Imbalance & Proposed Options for the US Response," dated September 18, 1977, outlined the principal circumstances leading to apparent Soviet strategic military superiority and described several essential US counteractions possible and necessary under recent self-imposed constraints. The paper was discussed with Olds and several congressmen on the House Armed Services Committee. The paper was also sent to AFSC, USAF-RDQ, the B-1 SPO, and Jack Trenholm.

Joe Brown was kept informed of my liaison with Congress as a private citizen no longer sponsored by the Air Force. However, he warned that he would have to fire me if IBM Headquarters became unfavorably aware of these activities. This had become a usual and acceptable risk.

My paper described the principal strategic circumstances expected during 1977–1987 including the evolving Soviet military posture, current status, improvements leading to their superiority in the mid-1980s, and their civil and aerospace defenses. These factors were compared to the US strategic military posture. The arms limitation accords of SAL I, Vladivostok (VLAD) and SALT II were also summarized with their attendant risks. The conclusions included the vital need for specific contingency developments in our Strategic Triad of forces._

Soviet Strategic Forces

As of mid-1977, the Soviets had deployed about 900 SLBMs in over 60 nuclear powered, missile-launching submarines. They possessed more than 1600 ICBMs of eight types in 18 variants. These missiles were armed with RVs, MRVs and MIRVs having up to 24 megaton warheads. The Soviet hardened counterforce target structure was much larger and far more hardened than those in the United States. Most of their counterforce targets were too hardened to be effectively attacked by our SLBM submarines.

The Soviet forces provided several times our readily deliverable strategic megatonnage and gave them an Equivalent Megatonnage (EMT) about twice that of the United States. They had deployed more than ten times the EMT needed to cause 150,000,000 casualties in the United States. Soviet Countermilitary Potential (CMP), their ability to destroy hardened US strategic targets (ICBM silos), was about equal to that of the US Triad, primarily due to the greater accuracy of our smaller forces.

The USSR began deployment of their fourth generation ICBMs in late 1974. These included the SS–17, SS–18, and SS–19 at a rate of 100 to 150 per year. They were also deploying three types of Delta SSBNs that would soon bring their SLBM force to the SAL I limit of 950 SLBMs.

A fifth generation then included four types of Soviet ICBMs under development. Their eventual deployment would expand their EMT to 20 times that needed to destroy half of the US population and all of our hardened strategic targets by the mid-1980s. The Soviets were winning a strategic arms race during the Carter Administration.

US Strategic Forces

As an effective deterrent force to offset the foregoing Soviet threats, the US had a triad of weapons consisting of: (1) 656 Polaris/Poseidon SLBMs in 41 SSBN submarines, (2) 1,054 ICBMs (1,000 Minutemen and 54 Titan IIs) in three variants having 2,154 RVs and MIRVs with yields from fractional to several megatons, (3) 500 operational bombers including 70 FB-111As and B-52 Models D, G & H. The US possessed a much smaller hardened strategic target structure than the USSR, primarily our ICBM silos and launch control centers.

The US SSBN/SLBM force had the best enduring pre-launch survivability for countervalue assured destruction but we would need to launch our ICBM forces when under confirmed attack (SLUCA). Various uses of the bomber force were described. Seven disadvantages that the US had accepted during the SALT I, VLAD and SALT II negotiations were also enumerated.

In spite of all the foregoing, the Carter Administration had, in just nine months, proposed much larger reductions to the 2,400 strategic weapon delivery systems and 1,320 MIRVed vehicles defined by VLAD. Further, they had unilaterally cancelled the Minuteman III ICBM and B-1 production programs and the deployment of ALCM A on the B-52s. They then set back MX ICBM development by several years. All this gave the Soviets little motivation to negotiate further strategic arms agreements.

Given the stance of President Carter and Secretary of Defense Brown, I recommended such actions as might be realizable during their tenure to mitigate the damage to the US future defense posture: (1) rapid deployment of the new Mark 12A MIRVs on the Minuteman ICBMs, (2) ensuring the development and deployment of the new Trident SSBN/SLBM Weapon Systems, (3) partial deployment of the ALCM A while awaiting development of the ALCM B for the B-52s, and (4) preserving the options for more rapid MX ICBM and B-1 Bomber development and production.

The only way that most of these vital strategic programs could be realized was through the defeat of President Carter by Ronald Reagan in 1980.

A Program Control Review of the B-52D Avionics Improvement Program at ESC-Owego on October 4 and 5, 1978, revealed major deficiencies in program management, financial controls, system engineering, and software development. This very large, fixed-price contract afforded FSD the greatest potential technical and financial risks of any program ever reviewed under MI 10-24. As a follow up, I conducted specialized reviews by FSD Systems Engineering with selected helpers and Finance.

Joe Brown insisted that I start a division-wide study to determine a three-year display strategy as part of IBM's overall strategy in displays and controls. The study was completed and I briefed it to John Jackson on August 1, 1978, and to IBM Vice President Branscom at IBM-Armonk on October 1.

I attended five classified national technical symposia in 1978 as part of the work for Joe Brown. These were usually pleasant but rarely informative. One in particular, in San Diego, provided information that in retrospect substantially affected the remainder of my professional life.

Ted Speaker from the Defense Intelligence Agency briefed us on the Projected Soviet Strategic Threat. Being very familiar with the topic, I quickly noticed a major change. Their fifth generation ICBM variants of the SS–17, SS–18 and SS–19 were now expected to have major accuracy and reliability improvements that had not been anticipated until the late 1980s. It was immediately apparent to me from the previous I/O-2 calculations and final report that these improvements would give the Soviets overwhelming strategic military superiority in the early 1980s instead of the mid-1980s!

Meeting with Ted Speaker after his presentation, I learned that he had heard about my prior work for Major General Keegan. Since note-taking was not allowed at the meeting, I requested permission to memorize several of his ICBM performance charts and ingested all the facts that could justify another study under Air Force

Invitational Orders. With Jack Trenholm retired and many Carter Administration weapon systems cancelled, it would take more time to find a sponsor.

Jack Trenholm had agreed in our telephone conversations that there was little to be gained by further dialogues with the Air Force during the Carter Administration. Very selective conversations were continued with the support staffs of the House and Senate Armed Services Committees. This evolved into discussions with Committee members Congressman McDonald and especially, Congressman Richard White (R-TX). Dialogues with staff members included Fred Smith, Captain Macan (USN), Mac Snowden, John Lolly, Peter Hughes and Rhett Dawson. Those men were generally Republicans and some were on the staff of the House Arms Subcommittee for Investigations.

Since these contacts involved high professional risks for an IBM employee, only formal requests for the September 1977 unclassified paper and the classified I/O-2 Final Report were fulfilled. The committee staffers found these papers to be very useful in their continuing efforts to cope with the Carter Administration.

While we all felt that President Carter's defense policies were damaging the future strategic defense posture of the United States, there was not much that could be done. These 1978 contacts and experiences on the Hill paid off in the early 1980s when the confidence that was established with these people helped me to be much more effective under the Reagan Administration.

On January 19, Jim Bitonti was "promoted out" of FSD. The senior managers, Cook, O'Malley and Driessen, jockeyed for position. With Bitonti gone, there was a slight chance that my professional fortunes might improve. Al Babbitt went to John Jackson on February 13 regarding a manager's position for me. Al later requested a 59-level job description for my current PCR and consulting responsibilities. Jackson was not receptive.

MX-C³ Red Team

I was still immersed in conducting my unpopular but effective PCRs during 1979. The Command, Control and Communications System for the planned MX ICBM deployment was a "must win" program for CSC-Gaithersburg. My formal PCR on their MX proposal effort during May 16–17 found 30 deficiencies; they were not competitive.

On July 26, a memorandum from Jackson directed me to form a "red team" to audit the CSC proposal effort on MX basing. As usual, on the more difficult tasks in FSD, I immediately enlisted John Cooney. Nick Kovalchick, a manager from the Owego years, also became an effective team member. By August 7, we had selected 14 people from many disciplines throughout FSD.

The Red Team studied the MX-C³ Request for Proposal (RFP) and all of the available formative CSC proposal documentation. We concluded in our three August meetings that the CSC proposal work was not a timely, competitive effort. On August 29–30 we held the first review at SDS-Manassas to cover their portion of the proposal.

The primary Red Team review of MX-C³ at CSC-Gaithersburg generated a running summary of all significant contributions agreed upon by the Red Team, CSC people, and the senior managers. The summary was completed about 10:00 p.m. on Friday, September 7. All participants, except for John and Nick and our faithful supporting staff, immediately and gratefully went home.

John had scheduled secretaries for the rest of the night and into the wee hours of Saturday, September 8. We dictated our report to the sequence of secretaries and took it to the MX-C^3 Proposal Manager's office building before breakfast. The door was locked; after all, it was Saturday.

We next went to FSD Headquarters to distribute our work to the senior managers involved in the review. Phil Whittaker requested that we brief and discuss our findings on Monday. After listening to our story, he told us to immediately inform the FSD Controller of the financial and manpower aspects of our report.

On September 13 John Jackson, recognizing that he had a big problem, announced that our boss, Al Babbitt, would become the MX-C^3 program manager if CSC won the competition. On September 18, we disbanded the Red Team, but Al's new role meant I had to stay involved in the proposal effort.

On September 19, Al Babbitt asked for a two-page summary of all five MX-C^3 volumes. He also needed a transmittal letter from John Jackson to the Air Force. Volume I was, as usual, the Executive Summary of the MX-C^3 Proposal. Al asked for a review the Summary Section. Since the text was useless, I rewrote the section entirely. He then requested a rewrite the other 80 percent of Volume I. John Cooney came back to the Headquarters for a couple of days and we split the task 40-40.

We both worked periodically until October 22, rewriting various volumes of the proposal for Al's approval. However, the proposal scope was too large for us to make many basic improvements. We at least kept the CSC proposal from being specifically unresponsive.

On December 11, the Air Force sent 117 questions to CSC regarding their proposal. These questions correlated very well with our critique in early September. CSC was unprepared and almost uninterested. Throughout the entire eight months, Vince Cook, the CSC general manager, was never present.

Many of the problems with the CSC proposal no longer had timely solutions. Bill McClain told us on December 15 that FSD Marketing was very worried about their prospects on MX-C^3. Their concerns were confirmed; CSC lost another big "must win" program.

Although I had worked diligently to help FSD win the MX-C^3 Program, I had severe reservations about the validity of the Air Force's Multiple Protective Shelter (MPS) basing scheme for the MX ICBM, given the Soviet ICBM capabilities. This apprehension was reinforced by the DIA intelligence briefing I had attended in September 1978. However, I had a year of very hard work ahead of me to verify these concerns.

Chapter 12. New Invitational Orders

During the prior six months, I was acutely aware of the imminent overwhelming Soviet strategic threat. With a lighter workload, I made inquiries at ASD-Dayton about obtaining a new set of Invitational Orders. I wanted to calculate the timing of the new Soviet threats in support of major counteracting improvements needed in the US defensive triad.

The B-1 SPO staff agreed that my earlier work for them should be up-dated. However, given the Carter Administration's hostility to the B-1 and the absence of Jack Trenholm to guide the efforts and take the heat from higher headquarters, they recommended a different arrangement this time. They deferred to Jack Cannon in ASD Plans. Jack was very knowledgeable about how the Air Force really functioned.

Jack Cannon and I had been giving each other a hand for 20 years. He called me on July 9 and said to contact Major Robert Reisinger at AFSC Headquarters, Andrews AFB, to arrange a meeting that he would endorse. This would allow indirect support from the B-1 SPO and move the work an echelon higher in the Air Force hierarchy, allowing me to work under the very broad charter of Air Force planning, and centering the efforts in the Washington, DC area.

I met with Major Reisinger, AFSC–XTRO, on July 17. We talked most of the afternoon about all aspects of US and Soviet strategic forces. I gave him the Secret I/O–1 & –2 reports and the unclassified summaries used in my congressional liaison.

Major Reisinger said that he would go to General Toomay and request my voluntary services as an independent consultant for the AFSC Vanguard Project, a current and major force structure planning study. All ASD and AFSC signatures for my I/O–3 studies were completed by December. In the meantime, the B-1 SPO issued interim orders that allowed my work to begin immediately.

My first task from Major Reisinger was to analyze a September letter from Secretary of Defense Harold Brown. He also provided the first of many Secret documents needed for the study. He said that an AFSC Intelligence Officer would later provide many more detailed documents.

November 15 was the first all-day meeting with Bob Reisinger. We reviewed many Secret documents provided by AFSC Intelligence and selected 47 that I could use in my FSD office. We also reviewed the written critique of Secretary Brown's letter on force structures.

Colonel Richard Wargowsky, Director of XTRO, was Major Reisinger's boss and another exceptionally fine Air Force officer. Dick and Bob were both command pilots with extensive combat experience in the Vietnam War.

On November 30, a call was received from Lt. Colonel Hal Gale, the newly assigned intelligence officer. Hal had a considerable sense of humor and noted that "intelligence officer" was a contradiction in terms. He had spent many years as an analyst at the AF Foreign Technology Division in Dayton and with a PhD, admitted to being a real rocket scientist.

During the first set of I/Os, much of the time was spent gathering secret data on Soviet forces from any cooperative source. Hal supplied carefully selected Secret documents faster than I could assimilate them.

On November 4, Jack Englund, now president of the ANSER Corporation, provided a complete update on the Soviet strategic target structure.

December 3 was the first meeting with my three Air Force supervisors, Bob, Dick and Hal. Reviewing my critique of Secretary Brown's letter, they all agreed that I had found a major error regarding survivability of the B-52/ALCM forces. The first test in building their confidence was successful. None of us could foresee that my work would later jeopardize the careers of these three officers.

The physical and intellectual effort of working 90-hour weeks from November 1979 to August 1980 took a toll. My sense that the Soviet Union would soon pose an overwhelming strategic military threat to the United States kept me going. This truth could only be established through a comprehensive and complex analysis of all US and Soviet strategic forces and their opposing target structures.

From 1958 through 1980, I had developed an analytical process to determine the strength and effectiveness of the United States and its chief rival. In the late 1950s it took only a day to evaluate the probable effectiveness of each bomber force against the opponent's strategic airbases and urban/industrial (U/I) complexes. By 1962, new mutual ICBM and SLBM deployments required more than a week of computations to determine the outcome of a possible general nuclear war.

The early 1960s were the so-called "missile gap" years. The Soviet ICBM deployments began to exceed those of the US. The Air Force put part of the B-52 force on airborne alert to mitigate this imbalance while the Minuteman I & II ICBMs were deployed in hardened silos throughout the central and northwestern United States. Mutual deterrence was thus well assured during the 1950s, 1960s and into the mid-1970s.

However, the USSR went on building ever greater numbers of increasingly capable strategic missiles. Now it took a month to calculate our relative strengths.

My first major report based on this effort, "Air Force Study of US and USSR Strategic Forces, 1975–1985," was prepared under Air Force Invitational Orders (I/O–I & I/O–II) during 1973–1977. At that time the data gathering and analysis took six months.

My 22 years of developing and performing these strategic force studies seemed to be preparation for one final such effort during 1979–1980. Brigadier General Delbert A. Jacobs, Deputy Chief of Staff for Plans and Programs at the Air Force Systems Command, wrote to John Jackson requesting an update of the foregoing study under

Invitational Orders (I/O–3). This was to be a contribution to the Air Force Vanguard master plan. John approved the arrangement for one year on a no-cost, non-interference basis, meaning that IBM would not seek reimbursement for this service and that my work on the project was not to cut into my normal IBM obligations.

Methodology

The Air Force I/O–3 Study was completed in August 1980. Getting government acceptance took longer — I was giving briefings from then until May of 1983. The new circumstances evolved since 1976 including more recent intelligence on Soviet strategic forces, the SALT II Arms Agreement, and interim improvements in comparable US forces. The first task in the new study was to integrate the best possible database on US and Soviet strategic forces. This threat synthesis for 1975–2000 required several months of gathering intelligence data from many sources. The progressive numbers, performance, reliability and availability of the SSBN/SLBM, ICBM and bomber forces of the current and projected US and Soviet triads were determined for each of the years covered by the study.

Each and all of the weapon systems of both sides were next characterized in terms of megatonnage (MT), equivalent megatonnage (EMT), counter-military potential (CMP) and later, prompt counter-military potential (PCMP). These factors determined their probable effectiveness against the major classes of hardened targets of the opposing side.

This process was followed by the traditional bean-counting of these progressively larger and more effective forces to show the relative strength of the US and Soviet forces during 1975–2000.

The Soviets had achieved essential parity by 1970. Thereafter, the Soviets began to deploy far more weapons and weapon systems than the United States.

Next, I compared the evolution of hardened strategic target systems on both sides. Counterforce targets included naval and bomber bases; ICBM silos and their launch control facilities (LCFs); command, control and communications centers; and nuclear weapon storage facilities. Softer targets included division-strength troop concentrations and major urban/industrial targets.

With this very large and time-progressive data base, it was then possible, although time-consuming and tedious, to calculate the probable results of an all-out, near-simultaneous counterforce exchange against the opposing target structures for each of the 25 years of the study interval, based on a thorough appreciation of how the sides would most likely commit their forces.

This analysis eventually established, beyond reasonable doubt, that the Soviets could destroy 95 percent of the US strategic targets in less than one hour by December 1981. Just by using their ICBM and on-station SSBN/SLBM forces they could accomplish an overwhelming destruction of our nation. They could also maintain this strategic superiority for the foreseeable future.

The first six months of the I/O effort required a dozen or more trips to AFSC Headquarters at Andrews AFB to assimilate the Soviet threat data provided by Lt. Colonel Hal Gale and Major Bob Reisinger. Bob also arranged two flights to SAC Headquarters during January and February, carefully timed to coincide with Omaha blizzards. We flew to Offutt AFB in the T-39, the Air Force version of the North American Saber jet. SAC was very helpful in providing numbers and performance data on their projected ICBMs and bombers. Our second trip to SAC on February 4 made the first flight seem relatively tame. Due to a severe crosswind, the T-39 landed even faster than normal and with less than 100-foot visibility. Commercial airliners would never attempt such a landing. As the only person of the six onboard who was not an Air Force Command Pilot, I noted that they didn't even look out the window or pause in their conversations. We stayed at SAC all week for very useful inputs to the preliminary Soviet and US force structure charts.

Threat Synthesis

By February 18, the synthesis was completed for the Soviet ICBM and SLBM forces through the 1980s. Overwork and stress from the implications of this work caused illnesses and nightmares. Extrapolation from prior studies indicated that the Soviet Union would establish overwhelming strategic military superiority within a year or two, primarily due to improved accuracy and more rapid deployments of their fifth generation of the SS-18 and SS-19 ICBMs.

Since the DOD and Air Force public and current classified positions did not reflect such a severe near-term Soviet threat, the summary charts were taken to AFSC for a meeting with Bob Reisinger and Dick Wargowsky on February 20. They provided copies of these charts to Major Siegel in the MX ICBM Project Office. A vigorous discussion followed regarding the possible vulnerability of the Multiple Protective Shelter (MPS) basing plan for the MX ICBM.

GOSG

Colonel Wargowsky and Major Reisinger needed time to verify the threat synthesis and figure out how to handle this "hot potato." Since their pro bono consultant was proving to be so productive, they asked for a review, too, of recent Secret papers on twenty topics that were normally provided throughout the Air Force by the General Officers Steering Group (GOSG).

I pondered 82 pages of GOSG papers, and then a Top Secret letter to the Congress from Secretary of Defense Harold Brown, in light of the new-found insights on Soviet forces, and reported on March 21. These analyses were similarly disquieting, too important to ignore, and terribly inconvenient. My sponsors worked out a disclaimer and letter of transmittal that distanced AFSC from the findings by essentially stating that this was just another private opinion on these vital issues. Brigadier General Jacobs then released the critique to other Air Force commands.

On May 15, the GOSG paper critique went to SAC with Major Reisinger and was sent through channels to ASD-Dayton. Colonel Wargowsky delivered a copy to Colonel John Conover at USAF Headquarters. Colonel Conover called on June 24 to say that he liked it. He added that my work on the GOSG papers had set them to re-examining the whole process of generating guidance papers.

I spent most of March 1980 at AFSC gathering and checking threat data. Hal Gale found an error in one of my summaries that required a weekend of calculations to correct. A set of ground rules was also developed that would be followed for both sides in a general war counterforce exchange when that phase of the analyses became necessary.

Preliminary Results

The fast-developing Soviet ICBM threat summarized in the charts was sufficiently alarming that the three of us met with General Jacobs on March 14. He asked to keep the charts for further study over the weekend. The preliminary work had produced far more information than they had anticipated. They spent the following week ensuring the validity of the Soviet ICBM charts with Hal Gale.

Throughout March and April, I went on to assess the Soviet SSBN/SLBMs and their Long Range Air Arm forces. By early May 1980, the entire Soviet threat synthesis was completed.

Using SAC data, the US strategic forces were similarly summarized for 1975–2000. To cover a spectrum of possibilities, I based the assessment on the most ambitious proposed deployment of 275 MX ICBMs in 6,325 multiple protective shelters (MPS), i.e., one missile shuffled between 23 semi-hardened MPS in the Carter Administration's scheme to ensure enduring survivability. The analysis eventually showed this maximum configuration to be entirely inadequate against even the conservatively projected Soviet ICBM threat.

Target Structures

The strategic targets in the United States were relatively easy to enumerate. Our airbases, naval bases, nuclear weapon storage sites, 1,054 ICBM silos and associated launch control facilities, and command, control and communication centers constituted fewer than 1,300 existing hardened targets. The SAC airbases were soft but time-urgent targets, bringing the total to less than 1,400. Major cities and industrial installations such as Hoover Dam brought the total list to about 1,500 strategic targets.

The proposed proliferation of MX/MPS installations could potentially add 6,325 targets hardened to 600 psi. Unfortunately, this initiative would be a decade too late. The analysis later showed that the Soviet ICBM deployments could destroy the MPS sites faster than we could build them, independent of their single MX missile for 23 shelters. Hardening the shelters to resist 300 or 1000 psi overpressures would make little difference in MPS survivability during a Soviet counterforce attack.

The Soviet target structure was an entirely different matter. Their much larger forces and fondness for reinforced concrete gave them more than twice as many targets and they were an average of three times the hardness of comparable US installations. Their greater population and huge defense industry also created over 300 strategic U/I targets that were not time-urgent.

Over the years, the most helpful source of intelligence data on Soviet targets was the ANSER Corporation, still run by Jack Englund. He provided much of the target data that I could not collect directly from SAC, DIA and AFSC through Lt. Colonel Gale. Jack was aware of the earlier critique of the GOSG papers, but the Air Staff did not respond to his request for a copy.

On Sunday, June 21, Jack spent a day in my office while the endless calculations continued. A copy of the critique happened to be on the desk. On the intellectual side of the defense business, effectiveness sometimes depends on a certain amount of "back-scratching" among thoroughly trusted friends.

Force Effectiveness

Having established and characterized the US and Soviet strategic forces over the 25-year study interval, the next step was to determine the effectiveness of every weapon; RV, MRV, MIRV, ASM and gravity bomb in the arsenal of both sides. As a first step, the opposing target system was delineated into classes of hardness such as 10,000 psi, 5000 psi, 3000 psi, 1000 psi and 300 psi overpressures that were necessary to destroy a specific hardened target. For perspective, normal atmospheric pressure is about 15 pounds per square inch.

Destruction of area targets, such as cities and airbases, required only 10 to 30 psi overpressure. Destruction of such targets would be limited only by the reliability of the delivery system. The yield and accuracy of almost any deliverable weapon, after 1970, from either side, would destroy a city or neutralize an airbase. In later years, each side had 10,000 or more readily available strategic nuclear weapons.

Nuclear weapon effectiveness against any specific target was primarily determined by the accuracy (Circular Error Probable, CEP) and megatons of TNT equivalent. For example, a one-megaton weapon delivered with an accuracy of 300 feet CEP would destroy just about any hardened target on earth or within 100 feet of the surface.

In the early years of the nuclear missile age, very high yield compensated for poor accuracy. The first large Soviet ICBM, the SS-9, had a delivery accuracy of around one mile, but a weapon yield estimated at 24 megatons. Such a weapon would have been burst at an altitude of 3000 feet or more to effectively neutralize airbases or destroy large cities, the primary missions of the early 1960s.

In the early 1980s, accuracy improvements due to better gyros and accelerometers in the guidance systems provided MIRV CEPs of 300 to 500 feet. This allowed yield reduction to 0.150 to 0.250 megatons while realizing the same effectiveness. Multiple MIRVs on each missile then increased the effectiveness of a single weapon system by at least an order of magnitude.

I calculated the probable effectiveness of each warhead against each of the various classes of opposing hardened targets for both the Soviets and the US in preparation for later use in the counterforce exchange. This included all of the weapons on all of the missiles and bombers over the 25-year study interval.

Clerical Comparisons

I calculated, charted and summarized the progressive increases in the US and USSR capabilities in deliverable megatons (MT), equivalent megatons (EMT), counter-military potential (CMP) and prompt counter-military potential (PCMP) — a term that was originated during the study to quantify the ability and effectiveness of ICBMs and SLBMs in the destruction of any target on earth within an hour. This was critical in counterforce warfare. Bomber forces were no longer a factor in the first hours of a general nuclear war.

Summary charts for the progressive MT, EMT, CMP and PCMP were completed for the end of each year in the 25-year study interval for both sides. The Soviets exceeded the US in total MT in the 1960s, and matched our counter-military potential in 1970 and our prompt CMP in 1976. Their current and projected strategic force deployments and improvements were designed to achieve greater superiority in all of these categories for the foreseeable future.

When all of the foregoing aspects of the study were concluded, I had just two more months to calculate the counterforce exchanges, finalize 40 to 50 illustrations, write and print the report, and a month for presentations and evaluations.

Counterforce Calculations

When the US/Soviet counterforce exchanges for 1974 and 1975 were recalculated based on the improved intelligence data on Soviet forces for 1976–1980, as expected, the results for the US were a little worse for those two years than reported in the I/O-2 report.

Differences in the numbers and capabilities of both sides then logically required an assumption that different tactics would be used in the commitment of force elements. Since the Soviets had quite reasonably and continuously emphasized ICBMs over their bomber forces before the US accepted these trends, they had much more to work with than the United States. The higher quality of the US force elements was not sufficient to offset Soviet superiority in throw-weight and numbers.

Logically, it would have to be assumed that the United States would launch its missile forces as soon as a Soviet attack was detected and confirmed — the SLUCA doctrine mentioned in the 1976 I/O study. Given the efficacy and variety of our radars and satellite warning systems, the US missiles would be launched within ten minutes after the first Soviet missiles were launched and detected. This would be 20 minutes before the impact of Soviet ICBM warheads. Similarly, the alert bomber and tanker forces would be scrambled from the SAC airbases.

The Soviet SS-18 and SS-19 ICBM forces were specifically designed to destroy the six US Minuteman ICBM fields that contained a total of 1,000 Minuteman Is, IIs, and eventually IIIs. Due to reliability and accuracy limitations, they would assign two warheads to each ICBM silo and associated LCF. The first would create an airburst, followed in a few seconds by a ground burst, timed to avoid fratricide due to radiation and debris from the first and nearby airbursts.

In the event that a Soviet ICBM failed to launch or exploded in flight before warhead separation, we expected they would promptly launch a back-up missile at the same target to ensure two-on-one target coverage.

Presidential Directive No. 59 from the Carter Administration asserted that there would be no US response before or during the initial Soviet attack. The remaining US forces were to retaliate with a later, devastating counterforce and countervalue attack. By 1980, my analysis raised the basic question — "With what?" Accordingly, the studies I prepared ignored this theosophical nonsense and were based on the simple premise of "launch them or lose them." In five years of briefing all of the I/O studies, not a single military officer objected to this premise, although it contradicted the Directive. Obviously, they (and fortunately, the Soviets) did not believe the US would behave in such an illogical manner.

US forces did not have such a complicated targeting scenario. There were simply not enough missile warheads for a two-on-one strike against a far more hardened and more numerous Soviet targets. Selective re-strikes would have to be performed hours later by the follow-on SAC bomber force. Softer targets could be attacked by the remaining SLBMs.

I calculated the Soviet potential counterforce attacks against the US during 1975–1992 in late June. The overwhelming Soviet superiority established by 1981 continued through 1992, in spite of any probable numbers of MX deployments, thus rendering it unnecessary to do specific counterforce computations for 1993–2000. The Soviet capabilities were reasonably projected to follow the trend during the final eight years of the study interval.

While preparing progressive illustrations of the study results, I went ahead with the continuous and interdependent calculations. This was still the best means of avoiding catastrophic errors while working alone during months of computation. Computers had been developed to handle this kind of number-crunching task a quarter century ago, but an adequate computer program to perform this work would have cost nearly $100,000 to develop. Further, writing the program would have taken more time than was available for the entire study. I used a 47-function scientific hand calculator for probability calculations and a desk calculator for more pedestrian work.

The story being developed was much too complicated to depict only in words. In early June, Don Amos, an excellent illustrator in the CSC-Gaithersburg Publishing Department, began to turn these lengthy charts into 40 or 50 properly scaled triple-page fold-out summary charts that covered the 25-year study interval. Then he was given nine more charts describing the Soviet counterforce attacks.

Bob Reisinger came over to the IBM office for the first time on June 13 to check my progress. On his second visit, July 3, he saw the charted results of the projected Soviet attack on the planned MX/MPS basing complex. He immediately foresaw a major Air Force reaction to the study.

Bob asked for a supplemental MPS hardness sensitivity study to determine if reasonably greater shelter hardness would significantly improve MX ICBM prelaunch survivability. A couple of days' work showed that there would be negligible improvement in survivability versus the Soviet MIRVs expected by the mid-1980s.

Bob wanted to see a draft of the final report by the end of July so that he and Hal Gale could determine the security classification. I was a few days late, but then, I was working for free.

End Game

By August 5, we had the final report. An ever-faithful and proficient typist, Marianna Wetzel, Earl Wells' secretary, had joined our marathon.

At 9:00 p.m., three draft copies of the report were delivered for Bob Reisinger, Dick Wargowsky and Hal Gale. At 9:00 a.m. on August 6, Bob Reisinger called from AFSC. He had not gone to bed and had spent the entire night reading the report. He generously said that the study was a "masterpiece." That alone made the effort worthwhile. The reception in some other Air Force circles would be much less enthusiastic.

Bob and Hal worked on the report classification for two days. On August 7, Bob came to the office at about noon. They had decided that the report could be classified Secret, since only secret data had been used to generate potentially Top Secret conclusions. No "new" intelligence data was generated in the process. However, the report contained a great quantity of intelligence data and some very provocative conclusions; therefore the classification would be Secret, WN Intel/FSRD. This meant that the report could not be declassified for 25 years, until 2005.

The very supportive publications staff at CSC-Gaithersburg gave me six finished copies of the 280-page I/O–3 Final Report, *Evolution and Capabilities of US and USSR Strategic Forces, 1975–2000,* with addenda that incorporated a special minor study on MX/MPS survivability and another that summarized all previous work(s) in support of the manned bomber and especially the B-1.

On August 12, one copy went to Al Babbitt, and four were delivered to AFSC Headquarters. Bob Reisinger had arranged for me to brief the study to senior staff members at AFSC. Dick Wargowsky, Hal Gale and various civilian intelligence people were part of the twenty or more in the selected, very professional audience. The success of this meeting gave me a brief sense of relief.

The next day Major Reisinger took one of the final copies of the report to the Air Force Ballistic Missile Division in Los Angeles.

Given my concurrent obligations to IBM, August 15 was my first day off after more than seven months. The last five weeks had been the most demanding. We took the rest of August as vacation while Bob Reisinger and Dick Wargowsky discussed the study within the Air Force. The report was well-received; field grade officers found it accurate and useful.

The generals would be another matter.

Audits & Presentations

On September 3, I had discussions with Bob, Dick and Hal at AFSC, then we met with Brigadier General Del Jacobs. It was clear that the findings would be very contentious if made public, even within the Air Force, during this election year. We all agreed that the provocative study should be closely held and thoroughly tested, but I pressed to have the Air Force either logically refute the conclusions or respond to the findings.

General Jacobs later directed Bob Reisinger to thoroughly audit the technical and mathematical aspects of the study and Hal Gale to again review the validity and use of all intelligence data. Bob spent five days at my office, walled in with six notebooks of the calculations and all of the draft illustrations. Everything checked out.

Bob knew that the analysis of MX/MPS survivability, generally less than three percent during 1986–1992, would be very ill-received by the Air Force upper echelons. He requested a separate verification all of the calculations related to this finding; all those calculations were also found to be valid.

September 18 and 19, Hal and Bob and I did another audit on the use of the intelligence data. Hal affirmed that all of the data had been used correctly and without distortions to support any particular conclusion.

The overall and specific validity of the report having been sustained, the next move was up to my determined and hapless sponsors.

On October 2, I was summoned, not invited, to meet with Colonel Al Casey, Deputy Chief of Staff for Systems Development at AFSC, and a Lt. Colonel, his Executive Officer. Colonel Casey forcefully pointed out several fallacious premises in the study including debris effects, targeting sequences, and in-flight failure detection. Sensing more trouble to come, I sat through this tirade quietly and made a mental note of each of his concerns without trying to defend the study. Colonel Casey had made an impression — but not the one that he had intended. He continued his successful AF career and later became a major general.

The AFSC Command briefing to four-star General Alton B. Slay and his executive council was scheduled for Monday, October 6. Obviously, this was going to be a tense meeting. Colonel Casey's several criticisms were unwarranted, but his challenge could not be ignored. I spent several hours on Sunday to prepare answers to his concerns.

The meeting began with encouraging words from my sponsors, then Colonel Casey suddenly supplied a one-page Secret document, unknown to the sponsors. He said that these questions would have to be answered during the briefing. A quick glance showed that the paper was a summary of his critique on October 2.

The audience in the large and secure AFSC auditorium consisted of General Slay, seven lesser generals, 40 or 50 colonels and other field grade officers, and a comparable number of civilians from the intelligence community, et al. After an introduction by Colonel Wargowsky, the presentation began with Vu-Graphs on a 20' x 30' screen.

At the earliest appropriate time, General Slay asked his first loaded question, and I provided a prepared and apparently passable answer. The second question was answered with a memorized three-minute response citing the page and paragraphs from a DOD higher authority report. The large room was very quiet. Like any smart general, Slay changed his tactics. His next interjection was not one of the questions for which I had prepared. He stated that the MX basing plan had recently been reviewed by a meeting of many four-star generals in the Air Force, and that they had determined that it was in the best interests of the Air Force. I replied that the study did not support their position and should be critically reviewed.

General Slay walked out.

Jacobs was no ordinary brigadier general, either. Instead of declaring the meeting over, he invited the audience, including some of the generals who out-ranked him, to stay and hear the rest of the presentation. Everyone stayed another hour, without further questions.

General Slay was disciplined in the use of his time. He seldom met with subordinates on any topic for more than 15 minutes. Following the briefing, he summoned General Jacobs for a very unpleasant two-hour meeting. He directed that the report never be transmitted directly from AFSC Headquarters, that all existing copies were to be taken to SAC Headquarters, and the next briefing was to be given there. Further, the presentation would only be given to people at the working level without any general officers in attendance.

The results of the study were provocative and General Slay was highly concerned that it had serious implications that exceeded the responsibilities of his command. The study scope had ventured into operational matters that were more properly addressed by SAC. However, he observed that the report would help SAC shift their positions on certain major issues — indicating that he believed the story, but could not afford to propagate the results as Commander of AFSC. Slay also said that he could not take the report to General Allen, then Air Force Chief of Staff, because of what amounted to contradictions to Allen's recent public statements.

Later on, someone in AFSC took the report to their Judge Advocate General to determine if the briefings could be stopped and all of the final reports destroyed. The JAG advised that since they had paid nothing for the work and the report, they could not stop further distribution or the briefings if they were conducted within DOD security regulations. AFSC had estimated that the ten months of work normally would have required three man-years and cost them $300,000 — if they could have found an organization that could and would perform the study in the first place.

At IBM, Al Babbitt was aware of my struggle to produce the I/O–3 report while still meeting all my obligations to the company. He liked the report, recognized the importance of the findings, and was very surprised by the reaction of General Slay. (At this time, we knew nothing of what Generals Slay and Jacobs had discussed.) We told John Jackson about the AFSC briefing and he offered to call General Slay and try to mitigate the situation. I reassured John that the Air Force, at all levels, understood that my work for them over the decades was always done as a private citizen and was never a reflection of any kind on IBM.

SAC

Major Reisinger had the SAC briefings scheduled for October 16 and said we would fly out to Offutt AFB on General Slay's personal C-135 (Boeing 707) the day before. When I mentioned this posh arrangement to Al Babbitt, he suggested that after we crossed the Mississippi River, the crew would push their civilian passenger out the door — without a parachute.

Colonel Wargowsky, Major Reisinger and three other colonels were on the C-135. The latter officers collectively were to report back to General Slay on how the briefings were received at SAC. AFSC's 45 copies of the final report were also on board.

After an hour en route, the Major privately advised me that both he and Colonel Wargowsky would be forced into retirement next week if I failed to prevail in the briefings to SAC. In this league, they played for keeps.

Bob also mentioned that Lt. Colonel Gale had been accused by his superiors of providing Top Secret rather than merely Secret data. However, Hal proved to them that the documents were all classified Secret. As in my 1976 effort, otherwise Top Secret conclusions had been developed through careful analysis and integration of many Secret documents.

During the morning session, I presented the study to about 40 field grade officers. Such large audiences were uncommon at SAC. Although the study results were contrary to SAC's position on several issues, there was no contention and the main argument was generally well received. After the briefing a young SAC major said he was working on a relevant report but had not yet developed a term to describe prompt counter-military potential (PCMP). Would it be all right if he used the term? A couple of weeks later an article appeared on the front page of the *Washington Post* using the term PCMP. Word had gotten around the Command, so the afternoon session included nearly 60 officers and many senior civilians. Several of the latter group had been friends of mine for many years. There were fewer questions from this clued-in and friendlier audience.

Separately, the officer responsible for new SAC aircraft programs said that the arguments for the B-1 Bomber offered nothing new to justify the aircraft. The SAC position at the time was that all of their FB-111 bombers should be converted to a designated FB-111E version. The major modification would include a whole new fuselage section to carry more weapons and new fuel tanks that would improve the FB-111's severe range limitations. This marginal new program was believed to be the only bomber-force improvement program that SAC could obtain after President Carter cancelled B-1 production.

I observed that after a quarter century of bomber advocacy and 18 years on the B-1, there was little more that I could say. Taking a longer view, should new political conditions permit, other options had to be considered — especially the imminence and magnitude of the Soviet threat. That aspect was indeed new and would justify the more powerful bomber force that could only be immediately realized with the already developed B-1.

The more provocative conclusion regarding MX ICBM survivability of only three percent during and after the deployment in 6,325 multiple protective shelters, was surprisingly well-received. My briefing also advised that the need for the MX was so crucial to US deterrence and counterforce capabilities that if they had to take $12 billion in useless concrete to get the missile deployed, they should deploy the MX on any such terms. They could always erect the missiles without any protection, as a message sent through Soviet satellite surveillance that the MX would be launched under the SLUCA doctrine, before it could be destroyed. The deep Minuteman silos were also vastly preferable alternatives to MPS basing.

Following this session, a group of SAC military and civilian participants privately advised me that they had suspected that MX/MPS survivability could be as low as 15 percent. However, since my threat synthesis was the best that they had seen, they were willing to concede that the analysis indicating three percent survivability was probably correct.

The informal SAC response was objective and favorable. They warned us that their formal written response, to be provided to AFSC, would be less supportive — but that we had prevailed with the most authoritative audiences in the Air Force. With their already distinguished Air Force careers thus reassured, Wargowsky and Reisinger bought the drinks that night; several, in fact.

We were to return on the regular SAC Friday afternoon flight, whatever our fortunes at the briefing. However, if there were not over 30 passengers or a general officer waiting to go, the flight would be deferred or cancelled. There were several SAC airmen there with their families but not enough to meet our quota.

Colonel Wargowsky was resourceful as usual and, fully aware of how the Air Force really functioned in such matters, he instructed me to look irritated and impatient. Then he went to the Operations Office, pointed to his consultant, and advised them that as General Slay's personal representative their visitor had the equivalent rank of a major general. This was an all-time rank achievement, if only for a few minutes. Thanks to Colonel Wargowsky and a tailwind, we landed at Andrews AFB two hours later.

Follow On

Bob Reisinger kept me abreast of the response to our SAC effort. On October 27, Colonel Chambers called my office from SAC Headquarters. We had discussed an audit and follow-on visit to Offutt AFB, but Chambers said that the report was acceptable without any further audits. The upcoming presidential election also deferred any immediate response to the study.

On October 28, I gave a summary three-hour briefing to John Jackson, Phil Whittaker, Bill McLain and Al Babbitt. We also discussed the implications to FSD and IBM from four additional unclassified charts that I had prepared.

In November, the AFSC sponsors and their weary consultant were all in recovery mode from the overwork and abnormal stress. The election of President Reagan on November 4 was seen as a very hopeful development. The I/O–3 report fortuitously provided a very supportive threat study, if any were needed, and at the right time.

Ambassador Watson

We were already confident by September 22, following the AFSC audits, that the study was valid and offered a better perspective on the imminent Soviet threat than other secret documents or Command positions. Thomas J. Watson, Jr., was now the Carter Administration's Ambassador to the Soviet Union. Since I had had professional discussions with him on several occasions over the past two decades, I sent a letter to Ambassador Watson on September 24, offering to brief him on this latest work during his next return visit to the United States.

I also noted that the study recently had been audited by the Air Force and that it "...indicated a far more imminent and enduring threat to the national entity and our vital interests than publicly or privately represented by our government." The Ambassador politely declined the offer on October 22, noting that he was "confident that he had been kept as well informed as possible by all of the government agencies involved." He was hardly alone in his more sanguine view of the strategic imbalance.

Reagan Transition

The Air Force did little in any formal way to make a serious response to the study, particularly given General Slay's attitude as AFSC Commander, combined with the usual turmoil of the transition in preparation for the Reagan presidency.

Aviation Week & Space Technology published a major article on November 17, 1980 regarding the rancor between competing factions in the transition team. They were variously known as "elitists" who favored detente and arms control, the "Gucci loafer boys" who were somewhere in the middle, and the "California primals" who favored immediate increases in the FY1981 Defense Budget of $25 billion to make quick fixes in US strategic nuclear weapons programs. The latter faction was aligned with Richard V. Allen, Reagan's long-term foreign policy advisor, and William Van Cleave, who headed the DOD Transition Team.

Since the I/O–3 Study had already jeopardized, and no doubt damaged, the careers of the AFSC Headquarters sponsors, they were in no position to propagate the findings of the report. After lengthy telephone discussions, Jack Trenholm again agreed on personal initiatives as in I/O–1 and I/O–2. The AFSC sponsors were privately advised of these intentions. They did not object.

Drumming Up Support

While I was able to proceed with government officials whom I knew from prior studies, some guidance was needed on how to enter the current mêlée in DOD. Bill Beecher, with his recent, high-level experience in government was again very helpful in sorting out the major players and how to approach them.

I had worked with Major General Block at various times since he was a captain in the B-1 SPO 17 years earlier, and now he was the only general officer on the Air Staff that I could approach privately. He was advised of how sensitive the generals were to the latest report. Iconoclastic as ever, Emil extended an invitation to come and talk. I also advised him of the planned initiatives at higher levels in DOD.

General David C. Jones had generally been appreciative of the work on the B-70 when he was a colonel, 21 years earlier, and later in 1976 as Air Force Chief of Staff regarding the first I/O report. In 1980, General Jones was the Chairman of the Joint Chiefs of Staff, the senior military officer reporting to the Secretary of Defense and independent advisor to the president. Only an indirect approach would be appropriate.

After several telephone calls, I delivered a copy of the I/O–3 study to a major general on General Jones' personal staff, noting a more imminent and severe Soviet strategic military threat than had been considered probable in 1976 and supporting many of the General's recent public statements and those of the SAC Commander regarding Carter's Presidential Directive 59 (misleadingly premised on an ability to survive the first strike). It took a month for him to check the validity of the author and report, but in the end the response from the Office of the Joint Chiefs of Staff was quite encouraging. They considered the study to be a useful worst-case analysis that raised intelligence concerns, agreed that the conclusions were logical, and said that all points were being considered. They also said that the report added depth to their analyses and that the document had been referred to the OSD/MX Basing Panel. Our dialog went on for three months.

Another copy of the latest I/O study was given to Andrew Marshall, still the Director of Net Assessment for DOD, with a cover letter describing the correlation of the 30 referenced intelligence documents and the 1978–1992 counterforce calculations.

With Al Babbitt as the manager, 1980 was my best year at IBM in the past quarter century. He understood my motivation and supported my Air Force work, and cleared my path by eliminating much of the routine work that I otherwise would have had to do. I also sought his advice on some of my Air Force activities.

Chapter 13. Reagan Transition Team

The first six months of briefings and discussions with Air Force and DOD officials had ensured the validity of the study and report developed under Air Force Invitational Orders during 1980. Air Force civilians and field grade officers readily accepted the controversial findings. But the general officers were usually unresponsive, due to long-held positions necessary for their professional survival under the expiring Carter Administration.

During the three-month liaison effort, the Chairman of the Joint Chiefs of Staff, DOD Director of Net Assessment, and Major General Block were all supportive and encouraging. The real problem was the many Air Force generals. The SAC formal response confirmed it on January 5, 1981. The SAC critique reflected the established Command positions regarding the FB-111E versus the B-1 Bomber, and the MPS/MX basing plans.

The obvious course of action that was still open was liaison on these issues with the civilians on the Reagan Transition Team that would soon hold high DOD and Air Force offices in the new administration.

Threats

Late in December 1980, I received a call one evening from an IBM acquaintance who was a former Air Force intelligence officer with the Foreign Technology Division at ASD-Dayton. He was apparently calling on behalf of some very disgruntled Air Force intelligence officials. He said that the report had harmed their careers. I dismissed the warning as a half-baked effort by some unhappy officials blowing off steam or trying to discourage further initiatives.

On January 6, I stopped to check the mailbox before leaving for the airport and an IBM PCR meeting in Rhode Island. There was a large brown envelope, without a return address, and with too much postage. The envelope did not contain a written message — just a live shotgun shell. My trip was immediately cancelled and the Bethesda police and IBM Security were notified. Mail deliveries to our residence and business addresses were blocked and traps approved for the office and residence telephones.

The telephone warning two weeks earlier came to mind immediately. The envelope had a cancellation zip code near Andrews AFB. The combination called for the threat to be reconsidered. The Bethesda police provided a drive-by every half-hour from 11:00 p.m. to 3:00 a.m. We used our outside lights to signal that we were still safe.

I took my .45 pistol to the office. Returning home at 4:00 p.m., I searched the woods and house, .45 in hand, looking for trouble. Then I called the family and told them that it was safe to come home. They were very cooperative and tolerant of this misadventure. If the intent of the unknown nemesis was merely to slow down the distribution and discussion of the I/O–3 Study, he was partially successful.

I began to consider another possibility. During the past three years, we had employed a Viet Nam veteran to make repairs around the house. His last job, in late 1980, was so poorly done that he was not paid and we re-did the work ourselves. Since he had clearly left his better nature (and some of his marbles) in Viet Nam, he was another potential source of the threat problem. His workshop was not far from Andrews AFB, perhaps an unfortunate coincidence. The police were asked to discontinue their surveillance. Our lives were back to normal by the end of January.

Liaison

On January 2, William R. Van Cleave, then head of the DOD Transition Team, agreed to a meeting to discuss Soviet missile CEPs and review the I/O–3 report. Since he had not yet been provided with a safe for classified documents, his copy of the report was delivered to the Transition Team offices on M Street on January 5. Dr. Van Cleave soon went to the White House National Security Council (NSC) to work with R.V. Allen, President Reagan's principal advisor on foreign and defense policies.

At the M Street offices, I had a similar discussion with Dr. William Schneider. He soon went to the new President's Office of Management and Budget (OMB). We met again a week later to discuss what might be done to enhance US strategic forces in response to the I/O–3 Report.

On January 13, I met with William H. Taft IV, great-grandson of the 27th president. He had replaced Van Cleave as head of the DOD Transition Team. He was very receptive to a 30-minute summary of the I/O–3 report and accepted a copy, in spite of having a severe cold and the flu. I suggested taking two months' IBM vacation time to work for him voluntarily; he did not accept my offer. He promised to study the report and discuss the findings with DOD and Air Force leaders. We had several telephone conversations during the next two months, particularly with regard to MPS basing for the MX ICBM.

Since former FSD President Bob Evans was a candidate for a high office in DDR&E, he was briefed on January 23; he insisted we meet with John Jackson on February 2, and he encouraged John to get the story to the new IBM Chairman, John Akers. Bob also asked for briefings to Norm Augustine and Gene Fubini, Chairman and Vice Chairman of the DOD Scientific Advisory Board, respectively.

On February 2, I had a half-hour meeting at M Street with the late Secretary-Designate of the Air Force, Verne Orr. The transition melee made it possible to arrange solo meetings with future senior government officials as they were still trying to grasp their immense new responsibilities without staff support. These discussions gave them a condensed and current appreciation of the Soviet threat that was otherwise unavailable to them. This meeting and others with new senior officials were very timely and would have been impossible a few weeks later.

These initial contacts went on to arrange many other meetings with senior officials of the new administration, including the Acting Assistant Secretary of the Air Force for Manpower, Reserves and Construction; General McCarthy and his staff who were responsible for MX ICBM basing; Dr. David Perrin, Director for Strategic Systems in OSD Program Assessment & Evaluation; C.E. Estes, Director for Strategic Policy for the OSD Undersecretary of Defense for Policy, Dr. Fred Ekle; and Dr. Richard De Lauer, OSD Undersecretary of Defense for Research and Engineering (DDR&E).

USAF Aeronautical Systems Division

The Transition Team liaison was given highest priority, but I also felt it would be important to brief associates at ASD-Dayton on the latest efforts in support of the B-1 Bomber. Since the Invitational Orders had expired and General Slay was certainly not of a mind to renew the arrangement, IBM or personal funds would have to pay for the trip. Al Babbitt found a way to cover the expenses.

Due to President Carter's cancellation of B-1 production, the B-1 SPO was in limbo. Working with Jack Cannon and his boss, S. Tremaine, ASD deputy for plans, we arranged a large, broad-based audience that included the B-1 SPO staff and many others. Dr. Tony Caccioppo, still chief scientist for the Foreign Technology Division, also arranged a briefing for his managers and senior staff. As in 1976, Dr. Cacciopo called later to say that FTD had conducted several analyses that confirmed the validity of the I/O–3 report.

Protocols

Throughout the 1980–1983 liaison with government officials that was necessary to obtain a full response to the 1980 I/O–3 Study, these initiatives were first discussed with capable and sympathetic Air Force sponsors. Lt. Colonel Reisinger frequently gave general advice and suggestions and any liaison with the Air Staff was guided by Colonel Wargowsky. Lt. Colonel Gale was very helpful regarding contacts with the intelligence community.

In response to General Slay's concerns, all references to AFSC Headquarters and XTRO sponsorship were deleted in the final report. Eventually, all of my 47 copies of the document were placed with high government officials and four "think tanks."

I prepared an unclassified and provocative summary of the I/O–3 Study for use in telephone calls or other discussions when arranging private briefings with senior officials. The principal generalities included: (1) Soviet strategic military superiority was now well established and would prevail into the 1990s; (2) the Soviet ICBM threat to the entire CONUS counterforce and countervalue target structures was imminent and overwhelming; (3) Presidential Directive No. 59 would be meaningless for the next several years; (4) the MPS deployment scheme for the MX ICBM was futile and fundamentally inadequate since immense and growing Soviet ICBM forces rendered position location uncertainty for the MX missiles to be immaterial. However, the MX was vital to the future strategic force posture of the United States; (5) steps would have to be taken immediately to increase our alert status and begin a measured, longer-term arms buildup; (6) SALT II was uniformly disadvantageous to the US and did not impose any intrinsic limitation on either adversary; (7) the study had been independently verified by many separate reviews and audits.

By mid-March, 1981, word of the study was quite widespread within the upper echelons of DOD and the Air Force. Briefings became easier to arrange.

Defense Science Board

Norm Augustine, Chairman of the DSB, was contacted on February 2; I offered to provide the study and/or a briefing to members of the Board. He had heard of the study, but as President of Martin-Marietta Corporation he thought he could have a conflict of interest. His company was heavily committed to the MPS basing for the MX ICBM. He asked that the briefing be given to the DSB Vice Chairman, Dr. Fubini.

Gene was briefed in the Pentagon on February 20. He found nothing specifically incorrect in the report, but neither did he acknowledge that the work was an original contribution. Apparently, the imminence of an overwhelming Soviet strategic military threat and consequent vulnerability of MPS/MX sites were by now well understood by the DOD leadership. Providing the I/O–3 Report to the Joint Chiefs of Staff and the DOD Director of Net Assessment two months earlier likely had contributed to their new enlightenment.

The election of President Reagan was also likely forcing some re-evaluation of prior Command positions. Within the military, the problem was still how to convince and move the generals to take responsive actions.

USAF Headquarters

In a more positive vein, Dr. Fubini suggested contacting the analytical side of the Air Staff, specifically Brigadier General R.A. Rosenberg. Dr. Cacciopo had also suggested calling Colonel John Friel, his subordinate. Accordingly, a letter was sent to the AF Assistant Chief of Staff for Studies & Analysis on February 23 with a copy to Colonel Friel, his Chief Analyst.

Colonel Wargowsky at AFSC indicated a change of attitude in the Air Staff; an old nemesis, Jake Welch, now a Major General and everyone's boss, was also contacted He arranged for a two-hour briefing to Colonel Chris Chambers, Director for Strategic Forces, on February 24. Colonels Friel and Chambers and Captain Rowell also attended. Rowell had attended my I/O–2 briefing to the B-1 SPO in 1976, when he was a new second lieutenant.

Since Air Force doctrine coined in the Carter Administration had imposed a framework on Colonel Chamber's analytical efforts, their analyses presumed that the Soviets would hold 30 percent of their ICBM forces in reserve after the first strike. This presumption, however illogical, conveniently deferred overwhelming Soviet strategic military superiority by several years.

The I/O–3 study showed that by December of the current year the Soviets would: (1) be able to destroy all of the strategic military and urban/industrial targets in the United State in less than one hour, (2) launch all of their counterforce ICBMs in the first strike and simultaneously launch SLBMs from their submarines operating off both the East and West Coasts, and (3) maintain reserves after their first strike that included several hundred ICBMs to be reloaded into newly vacated silos, the majority of their SLBM/SSBN submarine fleet that was not on station, and their entire Long Range Air Arm.

After the briefing, all of Colonel Chambers' concerns about the study were resolved; he took a copy for further analysis and observed that it had been "a very interesting afternoon."

W. H. Taft, IV

Fubini's observation regarding DOD inertia in response to the imminent threat was reinforced by another visit to William Taft, now the General Counsel for the Secretary of Defense. He had learned that there was general agreement with the study findings at high levels in DOD and the Air Force, but, "They're not going to do anything about it."

Taft concluded that everything possible had been done in DOD and obliquely advised I take the initiative elsewhere to trigger an appropriate government response. "Elsewhere" could only mean Congress!

OSD Panel

The MX basing and new bomber issues had now become important to DOD, the Air Force, Congress and the press corps. Consequently, the Office of the Secretary of Defense (OSD) formed a special panel to study these matters. The principal was Dr. S.L. Zieberg, an assistant secretary of defense concerned with strategic weapon systems, assisted by Dr. Marvin Atkins and Colonels Eibling and McDonald. This panel became known as the "Supergroup" and later included Admiral Kidd, USN, and Dr. Ted Gold from the Atomic Energy Commission (AEC).

During March and early April, the Supergroup was briefed and provided with copies of the I/O–3 Final Report. The rapport established with them regarding MX, the B-1 Bomber, and a possible small mobile ICBM caused Dr. Gold to later endorse the briefing to the Townes Committee.

Townes Committee

The slow and mixed response of the military and pressure from the Congress and press called forth the Reagan Administration's characteristic response to a complex problem — form another committee. Dr. Charles Townes, a world-famous scientist on the faculty of a California university, was selected in March by Secretary of Defense Casper Weinberger to be its chairman. The committee was to unanimously recommend a final basing plan for the MX ICBM. Dr. T.S. Gold, serving both OSD for atomic energy and the Atomic Energy Commission (AEC), was an important member of the Townes Committee Staff.

Ted Gold was familiar with the prior work with the Supergroup and an earlier meeting with Colonel Lewis, Director for Radiation in the Defense Nuclear Agency. We agreed that the Townes Committee would benefit substantially from the I/O–3 briefing on the Soviet threat, without still another opinion (mine) on how to best base the MX ICBM. This was easily agreed, since my liaison had already begun through congressional channels to promote basing the MX in the deepest, hardened Minuteman silos for the prompt counterforce mission, and deployment of a small, mobile ICBM as the only means to realize enduring land-based missile survivability.

On the morning of April 27, the Townes Committee met again in their huge, secure Pentagon conference room. The audience of 150 to 200 military and civilian officials included most of the high-level people working on the MX basing problem. With an audience of this size, they had two members of the committee sit up front to question the witness or briefer rather than open up to random questions. The two questioners were Mike Mays, a senior civilian for the AEC Lawrence Livermore Nuclear Laboratory, and a retired Air Force three-star general, Glen Kent — both were bright, experienced and opinionated.

I opened the presentation by noting that the purpose of the briefing was to explain why there had to be a Townes Committee and to provide a thoroughly-tested threat synthesis of the Soviet strategic forces anticipated during the next 19 years, not to float more opinions on how to base the MX ICBM. Early on, the two inquisitors asked their expected questions. They were quickly convinced that I had a very comprehensive and accurate knowledge of the threat, and soon backed off.

The presentation stressed the imminence of an overwhelming strategic threat, gave thorough coverage of the need for the B-1 Bomber in various roles, and especially emphasized the need for the MX deployment — while thoroughly discrediting the fallacious multiple protective shelter basing scheme. This testimony substantially reduced the Committee's possible recommendation of MPS basing for MX.

Unknown to me at the time, Dr. Henry F. Cooper was in the audience. He was then serving as Deputy Assistant Secretary of the Air Force for Strategic and Space Systems. He had been working to find a survivable basing plan for the MX ICBM, including the MPS method. Two years later, this turned out to have been a very fortunate happenstance.

Major General Jasper Welch was also in the audience. He readily perceived the briefing's probable effect on the Townes Committee. Jake was necessarily very much the man, among the Air Force's general officers, who was charged with maintaining their positions regarding MPS/MX basing and the FB-111E versus the B-1 Bomber. On leaving the conference room, General Welch and six of his colonels backed me up against the hallway wall like an errant cadet, and Jake made forceful assertions about my always acting contrary to the Air Force's established positions (he still rankled from 1976). His apparent unwillingness to consider the actual facts only reinforced my conviction that their determination to go ahead with MPS basing for the MX and the FB-111E Program was ill founded; I pressed on with advocating the silo-based MX and B-1 Bomber alternatives.

On July 1, Dr. Townes reported to Secretary of Defense Weinberger that he was unable to obtain a unanimous finding on MX basing. This briefing or other inputs had at least kept them from using MPS. The indecision also kept the door open to further initiatives.

DIA Liaison

On April 2, a call came in from USAF Colonel Lawrence E. Pence, chief of the Sun Stream Task Force, part of the highly classified efforts conducted by the Directorate for Collection Management of the Defense Intelligence Agency, requesting my briefing.

This liaison built sufficient confidence with Larry to ensure mutual support of his "dark" program during contacts with DOD, the NSC, and Congress. The efforts to start a small mobile ICBM development and production program for enduring CONUS survivability were compatible with and supportive of his responsibilities.

Final DOD Initiatives

Various sources advised that Secretary of Defense Weinberger was floundering on the MX basing problem and was seriously considering some preposterous alternatives. This was confirmed by the sequence of four failed attempts using the Air Staff, DOD Staff, the Supergroup, and finally the Townes Committee. In OMB they had considered him to be "Cap the Knife," but as Secretary of Defense, "Cap the Shovel."

There was never an attempt to brief Secretary Weinberger because he did not seem to have the ability and background necessary to understand and respond to the I/O-3 work. Instead, a letter was sent to Frank C. Carlucci, Deputy Secretary of Defense (and later, Secretary) and followed up with his assistants. These men included Brigadier General C.L. Powell, J.A. Goldsmith, Dr. Ted Gold and Vince Puritano (Carlucci's executive assistant). Later on, Puritano advised that a meeting with the Deputy Secretary was not necessary because Carlucci had read and fully understood the final report.

Civil Agencies

Colonel William B. Staples, who was responsible for defense programs in the Arms Control and Disarmament Agency (ACDA), was sent a copy of the final report on April 7. I got to know Bill much better a few years later when he was a senior civilian manager in ACDA.

Chapter 15. Senate Testimony & House Study

Having done everything possible in liaison with the Air Force and DOD, it was time to attempt briefings to the Senate Armed Services Committee. Several senators were actively concerned with the MPS/MX basing plan.

Senator Tower

Ronald R. Lehman was a Legislative Assistant to Senator John Tower (R-TX), Chairman of the Senate Armed Services Committee. Ron later arranged a briefing to himself and some of Senator Tower's staff, including John MacFarland and Jack Gaffney. They were all unreceptive or mildly hostile. Clearly, their leader was committed to the Air Force's preference for MPS/MX basing.

Senator Tower later placed these men in prominent positions in the new Reagan Administration. MacFarland, as National Security Advisor, supported an ineffective venture into Lebanon that killed more than 280 US Marines in a truck bombing of their barracks. Gaffney became a hard-right senior official in DOD. Lehman later served as an ambassador in the strategic arms control negotiations.

Senator Laxalt and Senator Garn

Nevada and Utah were the states that would be most burdened by the MPS deployment of the MX ICBM. Consequently, letters went to the senior staffers for Senators Paul Laxalt and Jake Garn. Senator Laxalt (R-NV) was the senator most influential with President Reagan. Senator Garn (R-Utah) had the largest and most active staff addressing their common MPS problem.

Samuel D. Ballanger, the senior staffer and Director of Legislation for Laxalt, immediately perceived the value of my report and provided it to Senator Laxalt for discussion with President Reagan.

Joseph T. Mayer, a contact on Senator Garn's professional staff, and several other staffers were quite interested in the report. He immediately gave me an outline of their own effort to assess alternative MX basing modes. Their staff study intended to examine nine criteria. Primarily a politically-oriented group, they had begun with costs, socio-economic/environmental impacts, and institutional constraints. My I/O–3 Study had already addressed their other six criteria to an extent far beyond their capabilities. Further, if the report's strategic arguments were accepted, there was no need to further explore their political concerns.

In short, Senators Laxalt and Garn had just been given a solution to their problem of getting the MPS basing scheme out of Nevada and Utah. Objective analysis had proven the MPS basing concept for the MX ICBM to be fatally flawed.

A briefing to Senator Garn went well, and Joe Mayer called on June 10 to schedule testimony to Senator Laxalt's Senate Subcommittee on Military Construction.

Testimony

At lunch before testifying, we were surprised to be joined by Air Force Lt. Colonel Louis T. Montoulli, an advanced ICBM system staff scientist. He was then a special assistant for MX in the Office of the AF Deputy Chief of Staff for Research, Development and Acquisition. He was accompanied by Major Victor D. Bras, an advanced ICBM staff officer assigned to the same office.

Earlier, I had been requested to work with Lt. Colonel Montoulli and with Jack Englund, President of the ANSER Corporation, to consider alternative basing plans for the MX ICBM. Lou's doctrinaire and organizational preference for MPS basing was very apparent from these previous meetings. Lou advised me that he had been invited over to hear (and critique) the testimony to Laxalt's Subcommittee. The senators clearly wanted some real-time insurance of the validity of this testimony. The declassified portion of the testimony is contained in pages 211–232, GPO Publication 82-990-0; *Senate Hearings on MX Missile Basing Mode Alternatives*, First Session of the 97th Congress. There had been prior testimony during June 16–18 by witnesses from the DIA, Air Force, ACDA, DOD, the Army and the Congressional Office of Technology Assessment (OTA).

The senators in attendance included Laxalt, Garn, and Ted Stevens (R-Alaska), who was Chairman of the Senate Subcommittee on DOD. Several members of the supporting staffs were present.

The extensive, successful testing of the study by 27 offices in DOD and the Air Force was noted and stated that neither IBM nor any of these organizations was being represented by my testimony and that the Air Force had not paid for this effort nor influenced the direction, scope or findings of the study.

A very recent check on Soviet developments during the past year (with Lt. Colonel Gale) indicated that they had made even more progress than we had expected. This made their current strategic military superiority greater and the situation somewhat worse than described in the study. The current Soviet ICBM threat also ensured that enduring land-based missile survivability could only be achieved by a new force of mobile, small ICBMs. The presentation would also show why the proposed SALT II Agreement was uniformly disadvantageous to the United States and inconsequential to either adversary.

The Soviet ICBM threat was described in detail. Some of the principal points emphasized in the presentation included: (1) Only 820 Soviet ICBMs in their current force of 1300 would be used to eliminate 95 percent of the military and I/U targets in the United States. (2) The Soviet SS-18 and SS-19 ICBM forces were specifically designed to destroy our six Minuteman II & III fields. (3) Soviet on-line megatonnage was usually three times greater than that of comparable US forces. (4) The Soviet on-line Triad of nuclear warheads outnumbered ours in 1980 and continued the trend thereafter. (5) Early Soviet emphasis on destroying hard targets (Minuteman silos) caused their prompt CMP to exceed ours in 1977 and had greatly increased since then. (6) The US could not absorb a first strike and respond with more than a nominal and marginal countervalue attack. A significant, belated counterforce attack by the US would be impossible. The Soviets would tell us when the war was over and that we lost. President Carter's PD-59 was a self-defeating policy.

The three Minuteman II Wings and the Minuteman IIIs at Warren, Grand Forks and Minot AF bases were reviewed in support of the current silo hardening program and the Minuteman/MK-12A Program that improved the accuracy and doubled the yield of the warheads. A phase-down of Wing V at Warren AFB was also shown as a means to accommodate the MX ICBM deployment under SALT II constraints.

The senators' primary interest was the detailed analysis of the vulnerability of MPS for MX ICBM basing. The calculations demonstrated that MX survivability would never be greater than three percent of the total force at any time during the planned 1985-1996 deployment. Only three percent, after spending ten years and $12 billion pouring concrete into 10,000 square miles of Nevada and Utah!

More detailed testimony supported the Minuteman silo basing for MX, production of the B-1 Bomber versus the FB-111E, production of the Trident D-5 SLBM, a program to develop and deploy a small, mobile land-based ICBM, possible disruptive ABM defenses of our ICBM fields; redundant and robust command, control and communication systems; and increased alert status for our Triad for the next several years to better deter any Soviet initiative to utilize their newly established overwhelming strategic military superiority.

Censors

Secret testimony is always censored by DOD to ensure that classified data is not inadvertently leaked in formal, public GPO publications. For example, the censors always eliminated the statements that MX survivability in MPS was only three percent. However, an informed analyst could readily deduce this number from the remaining testimony.

The Soviets thoroughly understood this. A representative from the Soviet Embassy was always first in line to pick up copies of congressional testimony. The real value of DOD censorship was massive deletion, such as 28 of my 29 Secret briefing charts. After the last meeting with W. H. Taft IV, an old axiom came to mind — if your want secret information to legally become public, brief Congress.

Censorship and/or whimsical classification had another advantage: elimination of references to embarrassing disclosures during congressional hearings. The embarrassment to the Air Force during this testimony, although not classified, was completely eliminated from the public record.

Lt. Colonel Montoulli

Lou was well-chosen as a defender of MPS basing. His testimony acknowledged the value of most of the work I had performed for the Air Force during the past two years, but his intention was to deflect or dilute the main conclusions. He described the work that we had been doing together with the ANSER Corporation to find a better way to base the MX within the existing Minuteman fields, including a possible and partially effective disruptive ABM defense of these bases. Then he implied that the Soviet ICBM threat synthesis I presented was very much the "worst case scenario," and that the Air Force did not accept this summary of the estimated threat for the Air Force planning studies or their current evaluation of the imbalance in strategic forces.

I pointed out that in 27 briefings to the Air Force, DOD and the Townes Committee, the threat synthesis and overall findings had been successfully defended. Furthermore, I presented Steve Ballanger, Senator Laxalt's principal aide, with a Secret intelligence document, signed by a colonel in AFSC Intelligence (Air Force Systems Command). The Air Force Ballistic Missile Division (BMD) was conducting a "Strategic Missile System–2000 Study" to determine the forces that they would need to develop and deploy over the next two decades. As the senior command, the AFSC was required to provide them with a design threat estimate for the next twenty years. What the AFSC had provided was copied straight out of Sections II & III of I/O–3 Final Report; their new Command Position on the Soviet missile threat. Section II of the report described the technical characteristics of all Soviet strategic ballistic missiles expected during 1975–2000 and Section III contained the synthesis of these threats, including my three-page summary fold-out that had been developed in 1980.

The amused expressions of the three Senators made this dialogue the favorable turning point of my testimony. But it took four more months of effort by those of us in Room 1223 to ensure the eventual demise of MPS basing for MX.

With the MPS basing issue essentially decided, everyone loosened up a bit, even Lou. The senators now broadened the discussion.

Senator Garn agreed on the study ground rule of strategic launch under confirmed attack (SLUCA). Comparing that to an ABM defense for Minuteman ICBM fields, I noted that: (1) even a partially effective, disruptive defense would cost $10 billion and (2) such a defense might only be able to defer the decision to launch under attack for less than an hour, and then with a considerably reduced ICBM force. Consequently, it would be much less costly and more rational to establish the SLUCA doctrine. Certainty of the impending attack would have to be reinforced by very robust and redundant warning and command-and-control networks.

The Air Force fundamentally needed the MX ICBM as a counterforce exchange weapon far more than a smaller, mobile ICBM. Then and later they opposed the latter missile, which was required to enhance enduring land-based survivability. The Air Force did not want the mobile missile to detract from the MX Program. After the Carter years, they were generally reluctant to change direction on anything, let alone ask for two new ICBMs.

Montoulli opposed the small mobile ICBM and objected that it would be unacceptable to transport 25,000 pounds of explosive propellant and the plutonium warhead on the highways. Senator Garn noted that that would be a small risk compared to the overall danger of an attack. I noted that the US had the largest road system in the world and 11,500,000 trucks that were comparable to those needed to transport and support the mobile ICBM. All of these mobile decoys were already in existence, at no cost to the government.

The warhead and missile would always be carried in separate vehicles and assembled on site. This missile would be deployed to ensure a survivable means of retaliatory attacks against soft, non-time-urgent countervalue (I/U) targets. Response time was not important.

Laxalt asked about the feasibility of basing MX ICBMs at sea. However, the MX was a mass counterforce weapon requiring that all missile launch and warhead arrival times in the Soviet Union have a timing accuracy of less than half a second. This would be impossible for a large sea-based force. The SLBMs on our SSBNs already provided our best systems for enduring survivability and destruction of relatively soft targets.

Asked about deterrence in the coming decade, I observed that the Soviets had spent a quarter of a trillion dollars more than the US in the past ten years and had realized an overwhelming strategic posture from their efforts. This would require a much higher alert status for our "triad" forces over the next five years. We had to recognize that we were now "second among equals."

Over the past five years, the USSR had deployed the most awesome destructive machine ever known to mankind. We would need a decade to duplicate their force posture, if we decided do it at all. However, I added, alerted to the new imbalance of strategic forces, the United States could maintain an adequate deterrence. The Soviets had deterred us, for all practical purposes, for 30 years while being militarily inferior.

Asked for my opinion on arms control talks with the Soviets, I advised Senator Laxalt that I had reviewed translations of the so-called Penkovsky Papers 15 years earlier. The Penkovsky Papers were Top Secret Soviet planning documents that Lt. Colonel Penkovsky had delivered to the British Embassy in Moscow. The tone and content of his documents as translated had given me the lasting impression that the Soviet State was basically malevolent. I added, "Over the longer term, we must negotiate with them, but so far, we can't be accused of being shrewd Yankee traders. The existing and proposed arms control agreements are uniformly disadvantageous to the United States and intrinsically inconsequential."

In a further dialogue regarding Henry Kissinger and Presidents Nixon, Ford and Carter, I gave opined that "President Carter renewed the (arms) negotiations consistent with his deep personal convictions. His somewhat simplistic theosophical preferences were gradually imposed, either by default or intent, on US foreign policy, economic policy and defense policy."

All this was going on against the backdrop of the deposing of the Shah of Iran, rampant inflation, cancellation of the B-1 and deferral of the B-52/ALCM and MX ICBM programs. The faulty MPS basing scheme had also evolved in response to President Carter's PD-59 requirement to absorb the first strike prior to our retaliation.

In closing, I encouraged them to define, at one-tenth of the MX cost, the development and deployment of a road-mobile ICBM that could have several hundred missiles fully operational prior to the MX IOC of 1986. Assuming that the calculations were correct, fewer than 100 MX warheads might survive a concerted Soviet attack. The road-mobile ICBM would give us four times that survivability in one third of the time and one tenth of the cost. My prior liaison with Colonel Larry Pence of the DIA had been essential to formulating this comparison.

A week later, Garn and Laxalt released a joint paper that was based primarily on my I/O–3 Study and testimony; it included the entire list of recommendations and their rationale; several other recommendations dealt with socio-economic and environmental impacts. The senators also recommended substantial R & D in ABM technology that included a possible space-based ABM defense. The paper was headlined in the June 29, 1981 *Defense Daily*. I felt that with this successful campaign I had made a significant contribution to national defense.

Senator Tower Reacts

John Tower, Chairman of the Senate Armed Services Committee, was very disappointed, that is, furious, at the revolt within his committee by Senators Garn and

Laxalt. Tower's concerns were reported in *Defense Daily* on June 30. He had wanted them to wait for the recommendations of the Townes Committee, expecting the committee to endorse his and the Air Force's preference for MPS basing. Tower also objected to any small, mobile ICBM development and deployment.

Tom Wicker, in the *New York Times* on June 30, picked up on the Senate's struggle over MPS/MX basing. The issue would not be fully decided for two more years.

The House

The effort to influence senators was my primary emphasis during April–June 1981. However, liaison was conducted with both sides of the Hill, initially through common agencies, the Library of Congress, and the Congressional Office for Technological Assessment (OTA).

I met with Dr. Peter Scharfman of and Ashton Carter of OTA twice in March for extensive discussions of the I/O–3 Final Report and MPS/MX basing vulnerability. Several similar meetings with Jonathan Medalia of the Library of Congress had the same intent. OTA was a primary resource for congressional committees and they frequently testified to both houses of Congress.

My briefings had some influence, but not enough. John H. Gibbons, Director of OTA, and Lionel Johns, the Assistant Director, caused OTA's report and testimony to Senator Laxalt's subcommittee to be both catholic and comprehensive. They discussed ten possible basing schemes for the MX ICBM that included MPS and silos. The premise that MX was a vitally necessary prompt counterforce weapon system to be launched under SLUCA doctrine negated all these hypothetical options. Without the thorough analytical work of the I/O–3 Study and later testimony to Laxalt's subcommittee, OTA's contribution would not have been of much help to the senators.

A direct initiative with the US House of Representatives began on April 13 with a formal letter and I/O–3 Final Report submitted to Fred A. Schatzman, a professional staff member of the House Armed Services Committee (HASC). Fred scheduled a joint half-day Saturday morning briefing on April 25 with about 20 staff members from both the Senate and House committees. I found House staffers generally easier to work with than Senate staff, perhaps due to their excessive workloads and the lesser two-year assured tenure of their congressmen.

Roy Jones was the Counsel on Oversight for the House Committee on Interior and Insular Affairs. He arranged a private meeting in May with Congressman John F. Seiberling (D-Ohio), Chairman of the Subcommittee on Public Lands and National Parks. The study was soon briefed to seven congressmen on the Interior committees, including Manuel Lujan, Jr. (R-NM) the ranking Republican, Don H. Clauson (R-CA), Don Young (R-Alaska) and Robert J. Lagomarsino (R-CA). My final and private two-hour briefing was to Richard B. Cheney (R-WY). He questioned me about the sources, methods and the prior successful testing of the report and listened intently to the two-hour briefing.

Congressman Cheney had a very good memory. Speaking on PBS-TV a decade later, when he was Secretary of Defense for President Bush-41, he used the punch line from the briefing. "The Soviet Union can destroy all of the strategic military and industrial targets in the United States in less than an hour." Two decades later, Cheney, serving with Presidents Ford, Bush-41 and Bush-43, remained a powerful, conservative and remarkably unpopular advocate for US national security interests.

Interior Committee Staff Study

Roy Jones was the senior legislative aide responsible for a report to guide the Interior Committee(s) and other congressmen in their decision to allow public lands to be used for MPS basing for the MX ICBM. Such deliberations were required for disposition of public lands of more than 5000 acres. Since MPS basing would affect over six million acres of public lands, this was an important decision for the Interior Committee(s).

I consulted with Roy for about three months as he prepared a report, *Basing the MX Missile*; GPO No. 82-710-0, August 1981, to prove that MPS basing did not merit the use of any public lands. This was the best way to be helpful to the House of Representatives. To brief the House Armed Services Committee, my ultimate goal, would require gaining the congressional staffers' confidence and endorsement.

End Game

The Interior Department's final report and the catastrophic findings regarding the MPS/MX basing mode were rapidly transmitted to the House and Senate leaderships, the White House, and DOD. Congressman Seiberling wrote a cover letter to Secretary of Defense Weinberger on June 22 emphasizing that he recommended the denial of public lands for MPS basing of the MX ICBM and supported the foregoing alternative programs. The new congressional position on the issue was another important defeat for MPS basing.

The *New York Times* and *Washington Post* gave prominent coverage to the Interior Committee's report and attendant implications on June 24 — another nail in the MPS coffin.

My longer term goal was to give testimony to the very powerful House Committee on Armed Services. Roy Jones guided and supported a meeting with Alan Chase, the professional staff member for research, development and procurement.

Congressman Richard C. White (D-TX) and his principal aide, Bill Chin, were also briefed on the new study. White was appreciative, and had been receptive to the earlier efforts to support the B-1 Bomber in 1976. He also suggested writing to the HASC Chairman, Melvin Price.

Alan Chase supported a briefing to HASC Staff Director John J. Ford, and other members of the staff. They were also satisfied that the story should be given to the full HASC. Committee Chairman Price provided me an opportunity to testify during the hearings on the balance and composition of US and USSR strategic forces early in February 1982, during the Second Session of the 97th Congress.

Think Tanks

The I/O–3 Study and its provocative findings also had to be gotten into the hands of the peripheral yet significant civilian defense intellectual community. Jack Englund of the ANSER Corporation, and the ANSER senior managers and professionals were briefed in early February.

Then there was Alexander Tachminji, still the vice president and general manager of the MITRE Corporation. About 150 people attended the briefing in McLean, VA. I knew Charles Zraket, the current Executive Vice President of MITRE, from the Defense Science Board meetings in San Diego. Charlie arranged a briefing at their Bedford, MA, location to nearly 200 MITRE professionals on August 3.

Thomas Watson, Jr.

Tom Watson, after retiring from IBM and as Ambassador to the Soviet Union in the Carter Administration, had established the Center for Foreign Policy Development at Brown University, his alma mater. The new director there, Mark Garrison, was his former deputy chief of mission (DCM) in Moscow. Mark invited me to a "Conference on Nuclear Strategy and the MX Decision" in September 1981, chaired by Tom Watson. The attendees included professionals from MIT, Brookings Institution, Hudson Institute, Harvard, RAND, Jonathan Medalia from the Library of Congress, and Ashton Carter from OTA. William Perry, former head of DDR&E for President Carter, also attended. Watson and US Senators Chase and Pell attended the evening dinners and group discussions.

Chapter 15. Reagan's Decisions, & Senate Testimony

Some success had attended my prior work with the Townes Committee, Senators Laxalt and Garn, and the House Interior Committee. Throughout the summer, I stayed in touch with the staffs of the US House of Representatives and the Senate. We were all waiting for some results from Senator Laxalt's dialogues with President Reagan.

I also contacted Bill Nelson and Paul Mason at Rockwell International about efforts to resuscitate the B-1 Program. To broaden our sources of information I called the office of the governor of Nevada, Robert List, where I spoke with his staffer for MPS/MX matters, Jerome Wiesner. He advised that Senator Laxalt was still using my I/O-3 study for his work with President Reagan.

President Reagan's Decisions

On Friday, October 2, a call came from a very elated Sam Ballanger, Senator Laxalt's legislative aide. "We've won! The President has cancelled the MPS basing plan! They are going to put 100 MX ICBMs in existing missile silos and buy 100 B-1 bombers. Weinberger doesn't know this yet and won't be told until a half-hour before the 6:00 p.m. press conference. Don't miss it!"

President Reagan announced his $180.3 billion program to build 100 MX missiles and 100 B-1 bombers to "close the window of vulnerability" that had been an issue during his 1980 presidential campaign. He cancelled all MPS basing plans in Utah and Nevada. He would build the Trident D-5 SLBM, improve military communications, and continue advanced development of the "stealth" bomber. Most of the rest of our objectives were also to be funded. The small mobile ICBM was not yet part of his package.

A barrage of questions came from the reporters. The President turned to the hapless Secretary Weinberger and said, "Cap will answer all of your technical questions."

We had won. The President had cancelled MPS basing in favor of silos for the MX ICBM and re-starting the B-1 Production Program. The public dialogue intensified immediately after President Reagan's announcement.

The Press and the Public React

Governor List was quoted in a local news bulletin an hour after President Reagan's speech: "A tremendous victory; a very long and very difficult chapter in Nevada history has been closed." The Nevada Cattleman's Association and Mormon Church agreed. The Duckwater Reservation Shoshone Indians took the afternoon off to celebrate. Opposition to MPS basing included 75 percent of the Nevada population.

The *Washington Post* gave President Reagan's decisions more than half of their front page and two more pages inside. The headline was "Reagan Asks for 100 MXs, 100 B-1 Bombers."

> A visibly upset Senator John Tower (R-Texas), Chairman of the Senate Armed Services Committee told reporters that he was gravely disappointed [furious] in the President's decisions, especially on the MX...seemingly stung by a lack of consultation on the final decision, Tower said that Reagan's decision appears to have been made within a small circle without coordination of the best technical and military experts.
>
> Rep. William L. Dickinson (R-Alabama), Ranking Republican on the House Armed Services Committee, also said he was disturbed on the MX. They pretty well negated all the experts we've heard from for the past five years testifying before our committee...a lot of people who put their professional careers on the line and provided a lot of facts. Dickinson was referring to Air Force officers and others who have consistently rejected other deployment schemes in favor of various shell games.

Amen. We knew them all, and had prevailed in spite of them. From the first publication of the I/O–3 study, it had taken 14 months for the US government to accept its thesis. This had been a long, tiresome trip.

Senate Majority Leader Howard H. Baker (R-TN) predicted that "The plan would be approved by Congress basically in the form it was presented." House Majority Leader James C. Wright (D-TX) supported the new MX plan but opposed the B-1. So did Les Aspin (D-WS). Senator Strom Thurmond (R-SC) called the plan a positive step. Senator Edward M. Kennedy (D-MA) praised the MX decision but said it was a mistake to revive the discredited and extravagant B-1 Bomber.

Senator Tower said he had already asked Senator John W. Warner (R-VA), Chairman of the Subcommittee on Strategic Forces, to conduct detailed hearings on President Reagan's plan. Senator Warner supported the proposals. This new hearings initiative was noteworthy.

A *New York Times* editorial on October 4 praised President Reagan's decision regarding MX ICBM basing and production. They severely criticized the 100-plane B-1 production program as too expensive, potentially obsolete and politically motivated. Even in these earlier days, the *Times* reporting was not unduly influenced by the facts.

The President's decision was the biggest news ever for the *Defense Daily*, a specialized defense publication. They devoted five full pages to the substance of his positions, accurately and without editorial comment.

Joint Chiefs

On October 6, the *Washington Post* ran a critical front-page story selectively quoting General David C. Jones, chairman of the joint chiefs of staff. In an extensive interview, he had stated that "The Chiefs were not consulted in advance on the details of the President's new strategic weapons program [and that] he had doubts about placing the first 36 MX missiles in newly hardened Titan missile silos." (We shared that view, for some then-classified reasons.)

General Jones stated that he would not fight the President's proposals and that he had been an unsuccessful advocate for MPS basing for the MX ICBM. Privately, General Jones may have been greatly relieved by President Reagan's decisions to rapidly modernize our strategic forces.

The politicians were much less sanguine. Senator Henry M. (Skip) Jackson (D-Wash.) and Congressman Joseph P. Addabbo were very critical of the MX and B-1 decisions, respectively. Senators Sam Nunn (D-GA), Barry Goldwater (R-AZ) and Henry F. Byrd, Jr. (I-VA) endorsed the new plans. However, Senator Goldwater warned Secretary Weinberger to stay away from the idea of putting the MX in an airplane — yet another of Weinberger's less than astute study objectives.

HASC

The chairman of the House Armed Services Committee, Melvin Price, attacked the President's plan in the *Washington Post* on October 7 and promised hearings on the MX basing matter. He was supportive of the B-1 decision. Samuel S. Stratton (D-NY) "regretted that Reagan rejected the mobile MX option recommended by the Joint Chiefs and his failure to consult them before announcing his six-year strategic plan." General Jones, then and later, continued to support the President's overall plan.

The mêlée of support and criticism of President Reagan's plan was fully reported in the *US News & World Report*, October 19, p. 35–36, along with other factors such as House Speaker Thomas P. (Tip) O'Neill's (D-MA) opinion that "The B-1 Bomber was obsolete, foolhardy and a waste of money." They observed:

> That sort of criticism, taken with other Pentagon reservations, spells big trouble for a plan that Ronald Reagan had expected would win wide support. As it turns out, the President is now facing the uncomfortable prospect of a bruising battle with his ideological soul mates.

The firestorm generated by President Reagan's decisions was anticipated. When the formal report of my testimony was published by the GPO, it was very useful background material for staffers in supporting their respective interests. However, I prepared an interim unclassified memorandum explaining the "Rationale for Further Support of President Reagan's Program for Renewal of US Strategic Military Forces." The final paragraph endorsed the B-1 Bomber variant. The aircraft had been developed at a cost of $4 billion during the past 20 years. Although the deployment had been delayed by five years due to the Carter Administration's cancellation, the B-1 was still the most effective penetrating bomber that we could confidently deploy in the 1980s and possibly the 1990s. The airframe design allowed space and payload for the subsystems needed for effective penetration of the Soviet air defenses, e.g., advanced ECM and later, lethal defense.

This latest paper was distributed to the staffers at the US House and Senate and selected DOD offices that would need a solid rationale to support the President's plans. On October 9, the memorandum was sent to Vince Puritano, executive assistant to Deputy Secretary of Defense Frank Carlucci.

Puritano thanked me on November 2. He had sent the memorandum to the DOD Office of Research and Engineering, and suggested I could contact Colonel Randy McDonald with "any further suggestions on the issue." After 15 months of effort, it was good to know that this was now "our program," rather than just that of a lone, unknown, unwelcome, unpaid strategic forces analyst.

Senate Testimony

When Tower ordered John W. Warner (R-Virginia) to hold hearings on President Reagan's decisions to renew America's strategic forces, the matter was finally reopened. Thanks to support, perhaps from Senators Garn and Laxalt or their staffers, I received a surprise notification on November 9 that I would be welcome to testify on November 13. Senator Warner had already held nine days of hearings with most of the senior civilians in DOD and the Air Force, and many admirals and generals.

In the democratic process, all sides must be heard. The CDI, Center for Defense Information, was (and is) a left-leaning lobbying group founded by Rear Admiral Gene R. LaRocque (Ret.). We were selected to testify at the same hearing, probably in part so my briefing would off-set that of the professional dove. Admiral LaRocque's CDI press release headlined, "Reagan Policies May Bring War." LaRocque later testified that the proposal for a small road-mobile ICBM was "ridiculous" and that we should stop funding the MX Missile and B-1 Bomber.

Hearings

The hearing in Room 212 of the Russell Senate Office Building was attended by Senators Warner, Thurmond (D-SC) and later Levin (D-MI). Staff aides were present for four other senators. Senator Tower was represented by Messrs. Frank Gaffney and Ron Lehman, two active non-admirers. Since the opening session was not classified, Lt. Colonel Bob Reisinger was notified at AFSC Headquarters. He attended in civilian clothes. We did not recognize each other during the hearings. Most of the civilian audience consisted of press and TV reporters.

The proceedings of Senator Warner's subcommittee are reported in *Hearings Before the Subcommittee on Strategic and Theater Nuclear Forces of the Committee on Armed Services*, United States Senate; Ninety-Seventh Congress, First Session, GPO Publication 88-729-0. The recorded testimony is on pages 405-411. All of the classified testimony during a later hour-long classified session with Senator Warner was deleted through our mutual agreement.

Former Secretary of Defense Harold Brown had written an editorial for the *Washington Post* that morning. Senator Warner included the article in the record, pages 391-394, "Wrong on the B-1, Wrong on the MX." His writing supported all of the failed policies and positions of the Carter Administration and was very critical of President Reagan's new strategic forces plan.

Admiral LaRocque's testimony was a collage of unsupported opinion and truth by assertion on preferred Navy programs and strategies. (We agreed on the need for the Trident D-5 SLBM, but for differing reasons.) He was generally against anything else that President Reagan had proposed.

When it was my turn to speak, I briefly reviewed the major findings of the I/O-3 Study, "Evolution and Capabilities of US and USSR Strategic Forces, 1975–2000," with a short talk in support of silo basing for the MX, the D-5 SLBM, higher alert status for our strategic forces, gradual achievement of strategic parity, disruptive ABM defense, the small mobile ICBM and the new B-1B variant of the B-1 Bomber, justified in the context of a newly-established Soviet strategic military superiority that ensured they could eliminate all US strategic counterforce and countervalue targets in less than an hour. In short, I asserted that there was no longer any place to hide. If the Soviets attacked, we in Washington would be dead in seven minutes. Out in Nebraska, they would be blind in 20 minutes, dead in 30 minutes, and their land (the Minuteman ICBM fields) would be useless for a century. Some 150 million Americans would die during and following their attack. As Senator Garn had observed during the June testimony, any public interface problem for a road-mobile ICBM was academic.

The unclassified testimony session was followed by a Secret briefing in support of the unclassified testimony. Since there was no projector available to display the charts, Senator Warner sat with me side by side for about an hour, leafing through the I/O–3 Study. Senator Warner noted that Senator Laxalt was already having the work published from his June 18 hearings.

Reactions

While Admiral LaRocque had talked first, and long, only my testimony was reported in the relatively objective *Defense Daily*, on the front pages, on November 16, 1981. The most important classified finding of the I/O–3 Study was now legally in the public domain.

The headline was "Congress Told Soviet Strategic Superiority Well Established. Soviets Could Wipe Out All US Targets in Less Than an Hour."

> Further, Lee Carpenter of the IBM Corporation [no disclaimer] told the Senate Armed Services Committee the threat is imminent and overwhelming. The Soviets, Carpenter said, can now eliminate essentially all US strategic military targets and all industrial/urban targets in less than an hour. The Soviets need only use their on-line ICBMs and currently on-station SLBMs without committing their long-range aircraft.
>
> Carpenter, a strategic weapons consultant to the Pentagon for 30 years, told the Strategic and Theater Nuclear Forces Subcommittee, chaired by Senator John Warner (R-VA) that it is no longer possible to base a large missile in fixed sites in the United States that cannot be defeated in less than an hour. The current and anticipated Soviet ICBM forces preclude enduring survivability of our ICBMs in fixed sites in the Continental United States — independent of the realizable hardness and/or numbers. Consequently, using the modified Titan II and Minuteman silos will allow early deployment with the least new construction.
>
> Calling President Reagan's recent decision to discard the MX-MPS basing concept and base the MX in modified Titan II and Minuteman silos astute, pragmatic and deserving of your further support, Carpenter said the hardening should be limited to about 2000 psi. He explained that while hardening is ultimately futile against a concerted Soviet ICBM attack, hardening the silos to 2000 psi would preclude attacks by SLBMs and special weapons effects.

The newspaper also reported support of the small road-mobile ICBM as the only means of enduring prelaunch survivability, and all the other initiatives on the list, including the B-1B Bomber variant.

During questioning at the President's October 2 press conference, Secretary Weinberger told reporters that some of the MX missiles would replace Titan ICBMs in their silos, reflecting his usual faltering approach to such matters. On October 8, Marvin Atkins, DOD Director of Offensive and Space Systems, backed an article hinting a shift from Titan to the Minuteman silos.

In the *Washington Post*, a headline on December 10, 1981, stated, "Air Force Reportedly Eyeing Minuteman III Silos for MX Basing." After their initial reaction to President Reagan's decision for silo basing, the Air Force sensibly decided to base the first 40 MX missiles in the deep Minuteman III silos in North Dakota and Wyoming.

The existing silos had the most modern command and control structure; each Minuteman III Field contained 150 or more ICBMs, allowing for expansion beyond the first 40 MXs; and each new MX Field would be small enough to permit disruptive ABM defense of the fields if decided upon in 1984. The first MX missiles could be installed in late 1985 and initial deployment completed in 1987.

Boeing

The Boeing Company lost 1,150 jobs because of MPS cancellation. Boeing senior management demanded an explanation. The Washington area manager for MX, Minuteman and Advanced ICBMs, Dick Donmoyer, contacted Alan Chase at the House Armed Services Committee, who referred him to me. Chase thought that, in the national interest, Boeing could be given some help on the MX basing problem.

Donmoyer first called the FSD President, John Jackson, who confirmed my responsibility for cancellation of MPS basing for MX. Donmoyer then called me, insisting on a meeting to discuss these activities and the rationale regarding MPS basing. We both knew Boeing was paying IBM Commercial several million dollars per month for computer leases. Any misstep would surely cost me my job.

Alan Chase hadn't realized that the Boeing contact could cause such a big problem. He called Boeing and told them that any meeting had better be constructive. Further, I was scheduled to give testimony to the full HASC in February 1982. (No one wanted the HASC to hear that my volunteer work for the 97th Congress had resulted in separation from IBM.)

A secure room and access to classified files was needed for any meeting with Boeing representatives regarding my work as a private citizen. FSD Legal Counsel, Bob Moore, decided we could meet in an IBM office if they were escorted to and from the FSD Headquarters front door and not invited to the IBM cafeteria for lunch. We discussed having Bob Moore attend the meeting as insurance to prevent a possible Boeing set-up, but this would appear to directly involve IBM in the activities of a private citizen. We would depend on Alan Chase and the pending testimony to the HASC to preclude such machinations.

Thus I met with three Boeing representatives on December 8, 1981. They included Dick Donmoyer; Dick Miller, Customer Requirements Manager for the MX and Minuteman Programs; and John Blaylock, Manager for Advanced Ballistic Missiles. The latter two came in from Seattle for our meeting. The man to be convinced was John Blaylock. He had an operations research background — he had been a road-mobile missile advocate for three or four years in both the SALT II negotiations and OSD, and had worked with Jim Wade in DDR&E. The meeting was initially rather tense but on a professional level they accepted the findings, and we ended with a constructive discussion of what to do next.

Donmoyer called on December 11 and provided a rundown on the reaction at Boeing. He and Dick Miller personally considered our meeting to be quite useful in posing questions to their Pentagon sponsors. He noted that a Colonel Adams in

Air Force Strategic Systems was familiar with the I/O–3 Study and was "extremely complimentary."

John Blaylock found that the study substantially reinforced his prior positions. From their own private analyses, Boeing knew that 200 MX missiles in 4500 MPSs would not provide survivability. In effect, the matter was closed. Martin-Denver had also lost about 600 jobs because of the MPS cancellation. However, Norm Augustine confirmed that he also agreed.

Boeing had gained over $2 billion in revenues on the B-1 and the B-52/SRAM programs during 1964–1980 due to the previously described internal FSD management fiascos. On balance, Boeing still gained far more from their host's intellectual initiatives than they lost on MPS basing.

Chapter 16. House Testimony & ICBM Basing

The contention over President Reagan's new defense programs continued unabated throughout 1982. As Senator Tower and many senior Air Force officials came to recognize my role in the cancellation of MPS basing for the MX ICBM and renewal of the B-1 Program, I became very unpopular with some of them.

The most significant annual aerospace industry symposium was held every January at the Naval War College at Monterey, CA, with high-level military officers and senior civilians from DOD as speakers and attendees. I had attended these events for years as an IBM employee, but this time my participation was more noticeable and included an unsolicited evaluation of my recent government liaisons.

Major General Nelson (Ret.), who sponsored my first I/Os in 1973–1976, said that at SAC General Levitt had asked him, "Who is [this Carpenter]? What keeps him going? Why doesn't he stop?" Nelson had to explain, "Carpenter thinks he is right and that his effort is the patriotic thing to do." Another SAC civilian said that he had enjoyed my October 6, 1980 briefing, but suggested that I wait at least another year before a return visit.

Major General Jasper Welch, a long-time nemesis, had a suprising smile and a handshake, noting that the Townes Committee incident was unimportant because the MPS cancellation had been "edicted." The General was silent at our next meeting in Washington—apparently by then Jake was better informed about how President Reagan's edict came about.

The most forceful response to the I/O–3 Study efforts came from Lt. General Glen Kent (Ret.) who had been one of the two inquisitors during testimony to the Townes Committee. When I ran into General Kent in the hotel lobby, he berated that effort both long and loud, generating considerable local interest. This confirmed my belief that the briefing had prevented the Air Force and DOD from railroading the MPS basing scheme through the committee in early 1981.

As usual, I kept my own counsel during General Kent's sincere evaluation of that testimony. When he ran out of steam, I noted: "General, this has been interesting, but it is raining and we have missed the bus to the seminar. Can I give you a lift to the campus?" He accepted.

HASC Testimony

After months of dialogue and delays in scheduling the hearings on Strategic Policy and the Modernization Program of the Administration, the Committee on Armed Services of the House of Representatives met in February and March 1982. I was the last scheduled witness.

It had been five months since their meeting with Secretary of Defense Weinberger — the HASC grilling that followed President Reagan's October 2 decisions. The interim had allowed considerable rethinking by the defense establishment. The February 23–25 witnesses included General Lew Allen, AF Chief of Staff; Lt. General Kelly Burke, DCS for R & D; Richard Burt, State Department Politico-Military Affairs; Dr. Edward Conrad, Defense Nuclear Agency; General B. Davis, SAC Commander; Dr. Richard De Lauer, DDR&E; Major General Forrest McCartney, Ballistic Missile Office, AFSC; Richard Perle, DOD International Security Policy; Major General Grayson Tate, Army Ballistic Missile Defense; and Dr. Charles Townes, Physics Professor at UC-Berkeley.

I attended some of the prior testimony to determine the positions and attitudes of those in attendance. DOD officials Richard Burt and Richard Perle testified with remarkable and thinly-veiled arrogance and seemed to raise the hackles of the leading congressmen. I would easily avoid such a mistake.

The Hearings on Military Posture and H.R. 5968, DOD Authorization for Appropriation for Fiscal 1983 before the Committee on Armed Services, House of Representatives, 97th Congress, Second Session, are reported in GPO Publication 92-526 and HASC No. 97-33. My unclassified and censored Secret testimony is included in the last 45 pages, 323-368.

I invited Lt. Colonel Bob Reisinger to attend the hearings again. I met with him on February 8 at AFSC to re-verify and update Soviet threat data for the occasion. The Soviet posture was consistent with that projected on our last visit with Lt. Colonel Hal Gale in April 1981.

The hearings, both open and classified, took most of the afternoon. Major contentious issues addressed included the necessity for Triad renewal, the fallacies of MX ICBM basing in MPS, and the need for a small, land-mobile ICBM for enduring survivability. I supported President Reagan's programs, noting that the Minuteman was as an excellent ICBM but possessed only one-sixth of the prompt countermilitary potential needed to achieve parity with the USSR. Further, the MX missiles in silos were urgently needed as our primary counterforce ICBM; we needed the small road-mobile ICBM for enduring survivability. One type of land-based ICBM could not perform both missions. The MX, at 190,000 pounds, was simply too large to be truly mobile. The emphasis was placed on the imminence of the overwhelming Soviet strategic threat and how the US could best react to the current situation.

During the later session, following the Secret presentation, an hour of specific classified answers could be given and the credibility of my responses to questions was greatly enhanced. Alan Chase, John Ford and Tony Battista all called the following week to say variously that the testimony, both impressive and alarming, now had a very high credibility with the HASC.

Ron Mann

Late in the year, I made contact with the office for presidential personnel to seek a position with the Reagan Administration. This began a year-long association with Ron Mann, a former intelligence officer who was now associate director of personnel for the President, and it evolved into an unofficial consulting role for the President's National Security Council (NSC). (Paid positions at the NSC were often filled by military personnel from the Pentagon, thereby keeping the NSC budget down.)

The I/O–3 Study was presented to a small NSC audience on February 15. Copies of the final report were provided to Ron Mann and Air Force Lt. Colonel Bob Linehard of the NSC. During his unusually long tenure, Linehard accompanied President Reagan to the Reykjavik Summit Meeting with Soviet Premier Gorbachev. After leaving the NSC, he was rapidly promoted to major general.

I mentioned that presidents are usually limited to two major initiatives in their first term, but Reagan had already presented four. The NSC staffers enthusiastically suggested that there were more to come; and they noted that the Soviets were already testing the new president in Latin America. If we did not act, they felt, there would be a "domino effect" in the region within six months.

We went over the entire list of prior liaison efforts and the probable results obtained. Ron Mann said they were considering two possible briefings and jobs for me: T.K. Jones in OSD-DDR&E, and the Federal Emergency Management Agency (FEMA). We agreed to try T.K. Jones but I rejected FEMA, an agency known to be a repository for political mediocrities.

T.K. Jones, OSD

Ron Mann set a briefing for March 23 in the Pentagon. Jones was the OSD deputy undersecretary for strategic and theater nuclear forces. Air Force Major General Voight, John Gardner and a few others were given an hour-long briefing. Jones suggested that I go back to work on the Hill, but I had already accomplished everything that could be done there from the outside. When he began to lecture me on the need for bombers to do BDA and re-strike missions, a concept that I had originated 20 years earlier, I knew this not the place for me.

Knowing Mann's intelligence background, I referred Colonel Larry Pence to him regarding Pence's Top Secret intelligence-gathering program. They got on well. Ron was also consulted regarding congressional liaison, congressional staff support, and press interviews.

MX Basing

The endless contention on how to base the MX ICBM continued throughout 1982 in spite of the president's October 2, 1981 decision. Too many powerful political, military and industrial people with bruised reputations would not allow the matter to be settled.

On January 1, 1982, the *Washington Post* described the interim and possible later basing concepts for the MX missiles. Their staff writer, Walter Pincus, stated that the first 40 MX would be based in existing Minuteman silos beginning in late 1986. This would save $1.5 billion versus Weinberger's October 2 statement putting them in Titan II silos.

Pincus also wrote that Secretary Weinberger had directed the Air Force to explore three additional modes for basing MX: (1) Putting them aboard a "continuous patrol aircraft," (2) Burying them 3000 feet in underground "Cidatels," or (3) Putting them in additional fixed or "deceptively based" silos and protecting them with an ABM system. The first scheme was beyond serious consideration; the second negated the primary role of the MX as a prompt counterforce weapon system; and the third covered many complex, expensive, years-consuming and futile combinations of hardened silos and various ABM defenses.

On February 10, the *Washington Post's* Michael Gettler article was headed "Joint Chiefs of Staff Offer Gloomy View of the Power Balance." With President Carter retired and the I/O–3 Study and testimony in circulation, there was no longer any requirement to sound unduly optimistic. The Chiefs reported that "Analyses project that a Soviet strike against US missile fields could destroy a major portion of the US ICBM force if the US chooses to ride out an attack before responding." The Soviets must have appreciated the inference of this public rejection of Carter's PD-59 and implied acceptance of a US Strategic Launch under Confirmed Attack (SLUCA) doctrine. The Ship of State was at last changing to a better course.

The defense trade press finally picked up on my March 1 HASC testimony. They headlined the advocacy of SLUCA and the development of a small, road-mobile ICBM as the ultimate solution to the prelaunch vulnerability problem. They also noted the objection to the airborne basing of MX, the ultimate futility of ABM defense of the missile fields, and/or deep underground basing.

They quoted two conclusions: (1) "Since current and projected Soviet ICBMs precluded survival of hardened silos in the CONUS, and current strategic policies imply the need to promptly destroy the Soviet strategic forces in the event of general war, the MX must immediately respond to a concerted attack by Soviet missiles, and (2) Because true mobility is the only effective means for achieving the prelaunch survivability of our Triad forces, we must initiate development of a small, road-mobile ICBM. A near-term development of this missile could be configured from current hardware and several hundred deployed at one-tenth of the cost and in one-third of the time required for MX in the previously planned MPS (i.e. by the mid-1980s). Eighty percent of this force could normally be deployed on existing US military installations."

Dense Pack

The third Weinberger study objective mentioned above proved to be the most troublesome. "Dense Pack" evolved from this option. This MX basing scheme involved 100 super-hardened silos resistant to overpressures up to 20,000 psi and spaced 1800 to 2000 feet apart within an area of less than 20 square miles in several shapes of distribution.

The objective of the Dense Pack configuration was deployment of 100 MX missiles in close proximity to increase the probability of fratricide by the near-simultaneous explosion of Soviet nuclear warheads and to provide a relatively small area to facilitate more effective ABM defense. As momentum for Dense Pack basing increased, including a special study by the Defense Science Board scheduled for completion in September, the plan became known as "Closely Spaced Basing" (CSB). All of this and more was reported in the May 14, 1982 *Aerospace Daily*, p. 73-74.

There are areas in the northwestern US where the land is capped by a 30-foot or greater thickness of hard solid rock. Placing the very hard silos in the softer soil beneath this hard-rock shield would improve the survivability of CSB by about 20 percent.

Unfortunately, the USSR had possessed the means to destroy any realizable CSB for the MX ICBM for 20 years. The Soviets always fielded the best ICBM force they could for fighting a nuclear war every day. They could not hope to destroy our numerous new Minuteman silos in the early 1960s, so they used their extant technology to build the Second Generation SS-9 ICBM. They deployed more than 300 of these huge missiles to destroy the Minuteman Launch Control Facilities (LCFs) and the relatively small number of Titan II ICBM silos with integral LCFs.

The SS-9 inertial guidance system provided accuracy (CEP) of over 3000 feet, hardly a hard-target weapon. However, the Soviet nuclear weapons technology produced a warhead for the SS-9 estimated to have a yield of 24 megatons or 24,000,000 tons of TNT. That nuclear weapon would dig a true crater having a depth of 750 to 800 feet and a diameter of about 6000 feet. The apparent crater would be about 4000 feet in diameter and the apparent depth 300 feet due to the fall-back of debris. The crater lip would be up to two miles in diameter and have heights of 40 to 200 feet.

The proposed super-hard (20,000 psi) silos, if not destroyed, would be buried underneath crater and crater lip debris. A superbly engineered missile and silo, even if somehow intact, would be of little use if lying on its side under a hundred feet of rubble.

The SS-9 would not have to be aimed at individual silos; only the epicenter(s) of the CSB missile field. The inherent inaccuracies of the SS-9s 3000-foot CEP would ensure adequate distribution over the entire field. Given the 24 megaton warhead in dry soil or rock, four MX silos would be in each crater and eight more buried under the crater lip. Thirty obsolescent SS-9 ICBMs could effectively destroy the 100 MX deployed in a CSB field. When the SS-9s were retired, they could be replaced by modern SS-18s equipped with similarly powerful single warheads.

The ABM defense problem for defending MX in CSB would be simpler but would remain insolvable due to a precursor "ladder-down" attack on the ABM site, followed in minutes by 20 to 30 large re-entry vehicles. Since accuracy was not critical when it came to such RVs, this would negate any value of a disruptive ABM defense.

Could the whole CSB scheme be that wrong? I called Air Force Colonel Roger Lewis in the Defense Nuclear Agency on May 14. Since we both knew the then-Secret accuracy and yield of the SS-9 RVs, and all of the foregoing large nuclear weapon effects were in the open literature, we could work the problem on the telephone. Colonel Lewis confirmed my analysis and said that he would look into Soviet capabilities for timing and fusing the warheads to avoid fratricide. He called the next day to assure me that this was within their capabilities. He encouraged me to somehow put an end to CSB.

Nearly two years of efforts to communicate with the senior military had shown there would be no constructive response to the analysis. The congress could not react in time to preclude CSB for the MX. The NSC was the only hope.

I called Ron Mann on May 18 and advised him of the CSB vulnerability. He said that Judge William P. Clark, the President's new National Security Advisor, was giving his first major public statement at Georgetown University's Center for Strategic and International Studies in four days. Following current advice from DOD, he intended to endorse the work being done on CSB. Therefore, Ron asked me to bring the analysis to the White House the next day. (The Old Executive Office Building, a big granite pile on the west side of the White House, provides offices for most of the White House workers. To dignify lesser errands and meetings, it too is commonly referred to as the White House in the Washington lexicon.) We reviewed my five-page analysis and drawings on May 19. He took it all to the NSC.

Michael Gettler of the *Washington Post* reported on Judge Clark's Georgetown University meeting on May 22 under the heading: "Option of Deploying MX in Older Silos Supported." The first paragraph stated:

> President Reagan's national security advisor, William P. Clark, yesterday made a strong pitch for keeping open the option of temporarily deploying new MX missiles in older missile silos.
>
> In his pitch yesterday to retain the option of putting MX in a limited number of Minuteman silos, Clark was also challenging Congress, which has cut $1.5 billion from the budget for the MX because Congress opposes placing missiles in existing silos and wants the Pentagon and White House to come up with a better answer.

Earlier on March 23, the Senate Armed Services Committee had disapproved this funding to cancel all work on basing the MX in Minuteman silos (thanks again to Senator Tower).

After the NSC confirmed the validity of my new paper on the severe vulnerability of Dense Pack or CSB, they may have given some very spirited guidance to DOD.

> The most favored permanent [basing] plan at the moment is called Dense Pack, which involves building tight clusters of missile silos for the MX and which could also mean a total break with the unratified SALT II accord with Moscow, which does not permit building new silos.
>
> But Clark said that Reagan also made it clear that until a permanent home is found for the missiles the Minuteman basing plan is an essential hedge. The MX Program is too important to allow the risk of technical, environmental or arms control debates to delay the introduction of the missile into the force.
>
> Clark said the MX, even in the old silos, would be a clear incentive for Moscow to negotiate, and that even in such silos the MX gains in survivability. This claim was not explained.

Dense Pack or CSB study activity in DOD soon faded away.

Chapter 17. Military, Industrial & Congressional Liaison

Concurrent with my testimony to the HASC and work for the NSC I was involved in other advocacy endeavors.

Army BMD

The facts that led to President Reagan's cancellation of MPS basing for the MX presented the Army Ballistic Missile Defense organization with a big problem. The ground-based air and space defense mission had always been an Army responsibility. The Army excelled in deploying very effective anti-aircraft artillery and surface-to-air missile defense systems, e.g., Nike and Hawk.

CONUS defense against a massive Soviet ballistic missile attack was then (and still is) their unsolved problem. The Nike Zeus ABM System deployed in North Dakota was ineffectual for defending the Minuteman fields. The Air Force pragmatically endorsed this installation knowing that: (1) Cities could not be defended against ICBMs and SLBMs with then-extant technology, (2) The Nike Zeus acquisition radar would supplement other attack warning systems, and (3) Zeus might absorb 20 or more Soviet warheads before being destroyed in a precursor, ladder-down attack on the site.

When the Carter Administration proposed MPS basing for MX, the Army soon followed by studying and developing the Low Altitude ABM Defense System (LoADS). The components of this system would be hidden in vacant shelters in the MX/MPS fields to provide a terminal defense against Soviet MIRVs. After several years of moving their primary ABM developments in this direction, the cancellation of MPS basing left the Army wondering what to do next in their ballistic missile defense (BMD) mission.

I understood their dilemma, having worked for the Army AAA Command on Nike thirty years earlier and subsequently on many ABM studies. In March, I met with Colonels Evans and Fischer, and gave them the I/O–3 briefing that explained the sources of their problem.

We had a constructive discussion. I suggested they contact Herman Kahn at his Hudson Institute to get some independent thinking on what the Army might sensibly do next in ballistic missile defense.

Hudson Institute

In spite of the Army's dismay with the I/O–3 Study, I contacted Herman Kahn on April 13. He had missed the earlier briefing to MITRE-Bedford, but knew of the considerable success of my I/O–3 efforts in Washington. He agreed to a briefing to him and his professionals on April 21.

When he had left the RAND Corporation to establish his own think tank, Herman had chosen a rural and very pleasant site near the Tappan Zee Bridge on the Hudson River. This allowed him to be near the New York and Washington news centers and to facilitate the hiring of political scientists and arms experts.

Herman gave me a copy of *Soviet Aerospace* dated November 16, 1981. The text, translated into Russian and then back into English, was an exact copy of my testimony to Senator Warner's subcommittee as reported in *Defense Daily*. Herman wryly observed that we had a common problem. The Soviets would read anything that we wrote, but it was hard to be heard in the United States.

I suggested that if the Army BMD had not yet asked the Hudson Institute to reconsider their ABM programs, he might call them. Herman Kahn died within a year of our last meeting. I later learned that Army BMD did have two study contracts with the Hudson Institute.

HASC

Congressman James Nelligan (R-PA) was very interested in my prior testimony to the House Armed Services Committee on March 1. On April 22, David Nathan called from his office to ask for a TV interview. This was a new and risky venture, but professional risk had become a way of life. A TV studio in Room B301 in the Rayburn House Office Building was available to congressmen for such interviews. So on May 4, after considerable coaching on TV techniques and some make-up, I gave Congressman Nelligan a 20-minute tape that emphasized the Soviet threat and what we should do about the situation.

Joe Meyer (on Senator Garn's staff), who was still covering MX basing for Senator Garn, invited me to discuss the "Dense Pack" paper with Bob Bell and Alex Glitsman from Senator Percy's staff. On May 7, I heard from Senator Percy, Chairman of the Senate Foreign Relations Committee. They did not accept my offer to testify, but agreed to include a brief statement it in the record of the hearings. The paper: "Evolution of US and USSR Strategic Force Balance and Resultant Perspectives on Arms Control Alternatives," appears in the Appendix to the SFRC hearings of April 29–May 13, published for the Second Session of the 97th Congress in GPO Publication 95-413.

The paper for the SFRC began with the usual six major findings regarding the Soviet threat, the deeply flawed MPS basing scheme for MX, and preferred weapon system deployments, and a rundown of the 20 months of briefings and successful testing of these findings. A brief history of how the strategic force imbalance had evolved noted,

> The massive Soviet SS-9 and SS-11 ICBM deployments during the late 1960s established essential strategic parity in 1970. Although deterred by their own common sense and the immense, surviving retaliatory potential of the US, So-

> viet forces could have inflicted 150,000,000 US casualties at any time during the past 12 years.
>
> Throughout the 1970s the Soviets continued an immense qualitative improvement program for their ICBM forces, concentrating on the accuracy, weapon yields, reliability and sufficient numbers to destroy our six Minuteman fields. Their 308 SS-18s and 300 SS-19s represent the most powerful and effective military force elements ever conceived and deployed by mankind.
>
> A few months into the Reagan Administration, the Secretary of Defense stated that the Soviet ICBM forces have the capability to destroy more than 90 percent of our Minuteman silos in a first strike. A year later the President announced that the Soviets had established strategic military superiority.
>
> Although our overall deterrence remains adequate, the Soviets have achieved a potentially significant military advantage. Their hardened counterforce target structure is about 50 percent more numerous and variously twice as hardened as our counterforce installations. Consequently, our Minuteman ICBM force would need six times their current prompt countermilitary potential to effect comparable destruction on the Soviet counterforce target structure."
>
> ...[C]onditional Soviet strategic military superiority currently exists and will endure into the 1990s. As the newly established second among equals we must solve this problem by three responses: (1) greater force readiness for interim deterrence during the 1980s, (2) renewal of our Triad to achieve and maintain full strategic parity by the early 1990s,, and (3) START negotiations to seriously reduce strategic armaments, giving first priority to eliminating the most effective and de-stabilizing force elements, all US and Soviet silo-based ICBMs. Should START fail, the US must eventually deploy more than 300 MX ICBMs to achieve parity in prompt countermilitary potential.

The specific needs were described to enhance interim deterrence and other arms limitations proposals. The latter included the Soviet-proposed "nuclear freeze" at present force levels to stop President Reagan's new initiatives and ensure their continuing dominance in strategic forces. The United States needed to "arm to parley" or the Soviets would not likely respond to the START initiative. As an example, the Pershing II IRBM and GLCM deployments were needed in Europe before the USSR would be motivated to reduce their theater SS-4, SS-5 and SS-20 forces.

On SALT II, it was commented that:

> The SALT I, Vladivostok and SALT II agreements were carefully crafted by the Soviets to be consistent with their technology, production capabilities and force structure planning. Consequently, they established strategic military superiority within these intrinsically inconsequential limitations. Our forces could have been similarly modernized, but we did not make the necessary economic sacrifices.
>
> Since the Soviets can now obliterate all of the important military and industrial/urban targets in the US within an hour, they have little basic need for quantitatively larger strategic forces. Their improvements will continue to be qualitative and generally consistent with the unratified SALT II Treaty. Accordingly, there is little to be gained from formal ratification by the US Senate. Further dialogue in the SALT II framework would merely delay any serious START negotiations.

In conclusion, it was noted that the President's START initiative would involve very difficult and prolonged negotiations. His offer on May 9, 1982 was an excellent basis for genuine arms reduction. Six reasons were given on why the President's ambitious and pragmatic START proposal deserved support by the Senate.

Boston Globe

For 18 years, Bill Beecher and I had continued our periodic discussions of military and foreign affairs. After leaving DOD, Bill became the *Globe's* diplomatic correspondent. I gave him a copy of the SFRC think piece on arms control. I generally did give Bill any of my unclassified work that might be of interest to him. He did a major article in the Op-Ed Page (19) of the *Boston Globe* on June 25, 1982. His article reported on the foregoing, but focused on the "less realistic proposal" portion of the paper, as selectively quoted below.

> [Carpenter would propose to] trade $2 billion in free American grain to the Soviets each year in exchange for agreed dismantling of 50 to 80 ICBMs by them annually. And he would prepare to manufacture — but not do so — the new silo-busting MX missile as insurance against failure in negotiations.
>
> Carpenter would continue the rapid development of the 10-warhead, accurate MX missile, modify some deep Minuteman silos to receive some of the MXs and actually build production facilities for the new weapon—but without moving into actual manufacture.
>
> By deferring production of the MX, he calculates that the US would save about $2 billion a year. He would offer to use this money to pay American farmers for grain which would be offered to the Soviet Union free.
>
> In return, the Soviets would be required to destroy 50 or more of their giant 10-warhead SS-18 missiles and silos each year for the six years it would take to eliminate that force.
>
> Subsequently, the Russians would have to eliminate 80 or more of their six-warhead SS-19s each year until there was parity in the number of Soviet ICBM warheads and American Minuteman warheads.
>
> Carpenter concedes it would take an historic change of heart for the Soviet leadership to seriously entertain such a far-out proposal But, he insists, by agreeing, they would move toward a position of greater mutual stability in a crisis situation while improving the standard of living of their people. Similarly, the United States would be better able to redress its strategic disadvantage, while improving farm incomes.

Twenty years later and after the collapse of the Soviet Union, Verne Wattawa, retired from the government and working for a consulting company, became manager of an effort to directly funnel US funds and heavy machinery into the Former Soviet Union (FSU) to help them de-activate portions of their ICBM forces. The FSU was simply too broke to finance their own disarmament. After two very difficult decades, his task was quite similar to my 1982 proposal, but without any benefit to American farmers.

Defense Week

Jack Cushman, a reporter for this defense industry publication, asked for an extensive interview on June 9. On June 14, 1982 *Defense Week* headlined a two-page article with "Launch MX on (Reliable) Warning." My I/O–3 positions on MX/MPS cancellation, Dense Pack, SLUCA, and improved early warning systems were now thoroughly in the public record.

All MX basing schemes were rejected except the President's planned interim deployment in Minuteman silos. This was the least expensive way to realize the greatest prompt countermilitary potential. The unique and necessary mission of the MX ICBM was emphasized as prompt and massive destruction of the Soviet ICBM fields and associated launch control facilities through SLUCA. Otherwise, there was no need for the missile. ABM defense of our ICBM fields was described as futile against a ladder-down attack. Further, the MX should not be defended, but immediately launched to destroy the Soviet counterforce, minutes after their attack was confirmed.

Townes II

Secretary of Defense Weinberger initiated yet another committee on MX basing alternatives. The chairman was again Dr. Charles H. Townes, under the auspices of the DOD Defense Science Board. We discussed our separate testimonies to the HASC, recent work to defeat Dense Pack, and his new effort to redefine MX basing. I sent Dr. Townes a letter that included censored testimony to the HASC and the paper written for Senator Percy's SFRC. Nothing further was heard about his second effort, which was later followed by a final major commission in early 1983.

NSC

In late December, Ron Mann asked for a visit to the OEOB to discuss two new things that I could do to help the NSC. The NSC wanted to study possible Soviet initiatives during the next five years. I helped him structure the study and select possible principals recommended by the intelligence community.

Mann's second request was more personally demanding. The NSC was considering the European GLCM and Pershing II IRBM deployments in response to Soviet SS-20s being established in Europe. The NSC wanted personal and direct assurance that our Pershing II IRBM would meet the specified mobility and other operational requirements.

Ron arranged a visit the Pershing II prime contractor, Martin-Orlando, on January 4, 1983; I was the only witness to a demonstration of the missile's mobility and probable subsequent performance. The Martin Company had an oval Belgian-block testing road surrounding a lake near their assembly plant. Accompanied by Martin's senior management, we witnessed the Pershing II Transporter-Erector-Launcher (TEL) and supporting equipment vans rapidly transit this extremely rough road for several minutes. I timed them as they sited and leveled the TEL, erected the missile and began the count-down for a simulated launch. I noted that the target coordinates indicated northern Alabama. With their macabre sense of humor, the operators had targeted the Army's Redstone Arsenal, the cognizant agency for the Pershing II Program.

The Martin general manager stood by me in his office while I called the NSC to confirm the successful Pershing IRBM demonstration. This was the same room where I was accused of disruptive behavior by an ESC-Owego marketeer two decades earlier.

Chapter 18. IBM Exodus

On January 8, 1982, the US government dropped its 13-year anti-trust action against IBM, but the damage was done. The litigation costs to IBM were estimated at nearly $100 million per year during 1969–1981, and the legal action had inhibited and distracted IBM Corporate management at a critical juncture in computer industry evolution — a major factor in ending 50 years of IBM as the United States' premier growth company.

The reduction of the FSD population still caused performance problems. My primary responsibility was to do the Program Control Reviews of new development programs, but managers were resisting the process more than ever. When John Jackson transferred to IBM Corporate Headquarters, he was replaced by Vincent N. Cook; this was not a positive development. I eventually learned that several fifth level FSD managers had asked for my services in line management positions, but Jackson had turned them all down.

My three-year effort under I/O–3 and subsequent government liaison had been successful and would soon be completed. There was no longer any rationale for staying in IBM, and I could again design an exit strategy.

Despite the damage done to their careers, the sponsors of my I/O-3 study, Lt. Colonel Reisinger and Colonel Wargowsky honored Barbara and me with a dinner at the Ft. McNair Officers' Club and presented me with a walnut plaque commemorating the final report and topped by an 18-inch plastic alligator in joking reference to General Alton B. Slay. Understandably, there was never any higher-level Air Force acknowledgement of my three-year successful effort to improve our strategic forces.

Under Vince Cook in FSD there was now much more risk from my work as a voluntary defense consultant. Rumors circulated. A fortunate visit as guest speaker to an IBM Marketing Seminar by John J. Ford, staff director for the HCAS where he generously endorsed my testimony to Congress eliminated the FSD professional threat. The IBM Vice President for the Washington area, Chuck McKittrick, was kept informed regarding my intended obligations to the Congress and NSC. This reduced IBM corporate concern about these errands.

Having stabilized my immediate professional future, I asked Vince Cook for a significant assignment as a line manager. In September, he offered me a position as IBM proposal manager for the STARS program, the Source Time and Attendance Recording System that was intended to cover the 700,000 employees of the US Postal Service. The total revenue was estimated at $137 million and the budget for proposal and pre-proposal activities was $4 million.

STARS Proposal Team

FSD senior managements had assigned the STARS pre-proposal work to our Cape Canaveral Facility (CCF), now under FSD-Houston. Corporate management had been interested in STARS since 1979. IBM-Boca Raton had been looking into developing a Series 1 computer that would be the Data Collection Device (DCD), the smallest and most numerous of the many computers required for STARS. After six months of pre-proposal efforts, the team had burned through $600,000 but had accomplished nothing of substance. IBM-STARS did not have a system design, proposal schedule, adequate staff or coherent direction, and employees from so many locations were involved that it looked like an episode from the Keystone Cops.

John Cooney agreed to be the STARS Deputy Proposal Manager. John would be a very capable replacement for me, in the event that should be necessary, and he could also later be a very good program manager. We made massive staffing changes, teamed up with other technology companies, simplified the reporting structure, and consolidated the proposal effort at Gaithersburg, MD.

The IBM National Accounts Division (NAD) project manager ignored repeated requests for a new ROI (Return on Investment, IBM's primary financial analytical tool to determine the profitability of new ventures). When we insisted, the resulting analysis showed several gross errors (about $36 million in all) and misleading conclusions.

When the official Request for Proposal was received from the US Postal Service on November 5, we were ready. But by then John Cooney had completed an accurate ROI on his own. The results showed very conditional and marginal profitability for IBM in the STARS Program. The final question was whether the IBM-Commercial product line would be competitive in responding to the still over-specified, four-level STARS configuration. Except at the top centralized level of computation requirements, existing IBM computers were either too small or too large.

We met with Gordon Meyers and NAD senior management at Gaithersburg December 7–10 for discussions of John's ROI. NAD/FSD senior management pondered the options, and on December 13, they cancelled the IBM proposal.

Three months of meetings with FSD Vice President Gerry Ebker did not produce a new assignment. On March 23, he simply said, "Vince Cook wants you out of the business."

There was a larger than ever severance package offered, including nearly five years' salary, intended to discourage the well-deserved Open Door action. After a night of numbed reflection, I decided to pragmatically accept the package. We signed the severance agreements the next day.

I still had an obligation to testify to the HASC on April 22, despite bouts of depression and health problems. However, the cumulative effects of three years of continuous overwork, the exceptional stress of working for the government as well as IBM, the STARS Proposal, and finally being fired after nearly 30 years as a faithful and productive IBM employee were too much. I was hit with double pneumonia, bronchitis, laryngitis, sinus infections and pleurisy and, as I later learned, a silent heart attack.

By November 1983, I was still struggling to recover. It was a time filled with angst, job applications, interviews, and many disappointments. I gave priority to the intelligence community, DOD and ACDA. Ron Mann, still associate director for presidential personnel, suggested contacting Ambassador Ken Adelman, Director of the US Arms Control and Disarmament Agency (ACDA). Work in that area would draw on my expertise and interest, and would be consistent with my belief that strategic arms control agreements with the Soviets were essential to national security.

I asked many of my more recent associates in government to follow up with Adelman, including Jim Woolsey and Gene Fubini. With support from several congressmen and many other government officials, I finally landed a first ACDA interview with Michael A. Guhin on July 13. He requested several classified documents and arranged a meeting with Dr. Henry F. Cooper, who remembered my Townes Committee briefing. Hank's appointment as a US Ambassador was confirmed the Senate on November 1. He called on November 4, saying that ACDA needed a way to compare the relative value of ICBMs, SLBMs, bombers and cruise missiles in arms negotiations. Ever hopeful, I started work on the analysis and we met for two hours on the following day. Finally, I met with Victor E. Alessi on November 22. We got on very well. Vic intended to hire me as a GS-15-10 physical science officer in the Strategic Affairs Division of ACDA. This Class B Appointment had a non-renewable four-year term.

On November 25, Vic called with the good news: I had a position in ACDA!

In the meanwhile, Barbara had gone back to school. She completed three degrees in seven years, landed a job she wanted despite the difficult employment environment, and completed her MSW. We managed to take a few desperately needed vacations during the summer; but by the end of the year, I was in tatters. All the stress had been manifesting itself for months in the usual ailments: severe chest pains from prior shingles attacks, giant hives and scintillating sarcoma.

Chapter 19. The President's Commission on Strategic Forces & Testimony to the House

Fifteen months after President Reagan's decision on a renewed strategic force structure, there remained a very substantial and effective resistance to his new initiatives. His indecisive secretary of defense, Casper Weinberger, Senator John Tower, chairman of the Senate Armed Services Committee, and the senior professional military people were collectively unresponsive or actively resisting President Reagan's strategic forces renewal plans.

Four study groups and committees had been formed and failed to effectively support the President's initiatives: the first Air Force study group, the OSD "Supergroup" and two Townes committees. On January 3, 1983, Reagan elevated the status of this effort by forming "The President's Commission on Strategic Forces" (PCSF). The Commission was chaired by Lt. General Brent Scowcroft (Ret.) and was informally known as the Scowcroft Commission.

Many of the Commission members had participated in the prior study efforts. Some of the ten selected members were John Deutch, Alexander Haig, Richard Helms, William Perry, Thomas Reed and James Woolsey. The seven senior counselors to the Commission included Harold Brown, Lloyd Cutler, Henry Kissinger, Melvin Laird, John McCone, Donald Rumsfeld and James Schlesinger. Marvin Atkins was the executive secretary.

Jim Woolsey remembered the effectiveness of my I/O–3 Study briefing to the first Townes Committee, and asked me write an unclassified paper for the Commission, based on the study and my subsequent testimony to the Congress, to provide the Commission with a focal point for their deliberations.

I had advocated the small, road-mobile ICBM throughout the study and government liaisons. This was the only program outside of the President's strategic initiatives. Since Air Force Colonel Larry Pence had provided me with data and insights after the I/O–3 Study was published, I had helped him contact Congress, Ron Mann, the NSC, and Jim Woolsey with regard to his Top Secret Project Sun Stream. Our efforts were very complementary regarding the small ICBM, so he agreed to critique that aspect of the new paper.

I delivered "Capabilities of US and USSR Strategic Military Forces and Consequent Constraints on the Evolution of US ICBM Forces" to the Commission's Pentagon office on January 19, including those for General Scowcroft, Jim Woolsey and Marv Atkins.

The paper, in summary, described the US–Soviet strategic force imbalance and the major policy and basing decisions required for modernization of our ICBM forces. Two types of ICBMs were supported, MX for the prompt counterforce mission and a small mobile ICBM for enduring survivability in the CONUS. Constraints on MX basing options and the vulnerability of MPS and closely spaced basing were described in support of MX installation in Minuteman silos. The MX Program was fundamental to regaining strategic parity and successful START negotiations. A small mobile ICBM was proposed as essential for stable CONUS-based survivable deterrence and overall reductions in US and USSR silo-based ICBM forces.

The paper in particular advocated renewal of essential strategic military parity through enhanced interim deterrence, Triad improvements, and START negotiations. Since the Commission would focus primarily on ballistic missile forces, the unique and necessary mission for the MX ICBM and the basing options were defined. The 1981–1982 studies were summarized and I posited that the Soviet reactive threats to Closely Spaced Basing (CSB) would probably be large weapons first, and later, earth-penetrating warheads.

The President's best and most pragmatic interim option was MX-basing in the Minuteman silos. With true mobility as the only means for achieving enduring prelaunch survivability in the CONUS, I also outlined the development, operational concept, utility and possible Soviet reaction to the small mobile ICBM. The paper's summary and recommendations were:

> Beyond a few hundred US and USSR ICBMs needed for mutual deterrence, these forces are destabilizing and configured primarily for counterforce warfare. The additional ICBMs are unnecessary for rational deterrence and wasteful of our mutual resources. Our long-term objective must be equitable arms reduction to achieve parity at the lowest practical level of these armaments that is consistent with mutual security.
>
> If START negotiations fail, we must build the MX force to achieve parity in prompt counter-military potential and try again to negotiate lower ICBM force levels. These arduous and enduring efforts to minimize the probability of conflict can best be supported by the following measures:

Maintain exceptional diligence and periodically higher alert rates for the bomber and sea-based deterrent forces to offset Soviet conditional superiority in the 1980s.

Ensure reliable and redundant early warning of any en route attack on our ICBM fields and the robust command, control and communications needed for timely launch of these missiles against Soviet counterforce targets.

Major Soviet aggressive initiatives will continue to be deterred by these precautions, ensuring both national security and the time necessary for measured, non-provocative modernization of our strategic forces and realization of significant arms reductions.

All of the excessively imaginative and expensive MX basing schemes attempting to provide enduring survivability should be rejected as illogical, unnecessary and ultimately futile.

Continue the development and flight testing of MX, establish production facilities, and modify many of the deep Minuteman silos for MX deployment. This will provide the overdue modernization of our ICBM forces, bargaining factors needed for START negotiations, and the means to regain strategic force parity with or without success in START.

Produce and deploy MX in Minuteman silos only to the extent needed for modernization and to regain parity through START negotiations at much lower levels of US and Soviet silo-based ICBMs.

Develop, test and deploy the small road-mobile ICBM as the only means to realize enduring prelaunch survivability in the CONUS.

When fully developed in the late 1980s, the small mobile ICBM would provide relative invulnerability and non-time-urgent retaliatory capability against Soviet counterforce and countervalue targets. This would mitigate the requirement of higher alert rates for our strategic aircraft and SSBNs. Mutual deployment of the road-mobile missile would also facilitate eventual elimination of destabilizing US and USSR silo-based ICBMs through later phases of the START negotiations.

Jim Woolsey agreed that it would be desirable later to distribute the PCSF paper to congressional staffers and selected military people who had found such works to be useful in the past. In the first three months, 15 copies were given to Alan Chase and John Ford on the HASC staff, Ron Mann and the NSC, Colonel John Conover, USAF DDR&E; and Colonel Lou Montoulli, now on the White House Staff. On the Senate side, Sam Ballanger was copied for Senator Laxalt, John Meyer for Senator Garn, Carl Ford for the SFRC and Frank Gafney on the SASC Staff.

HASC

After seeing my PCSF paper, the House Armed Services Committee staff asked me to take a look at a new initiative of theirs that was endorsed by several congressmen, led by the Minority member William Dickinson (R-AL). The new initiative was a briefing to the Scowcroft Commission that offered congressional support for 100 MX ICBMs in the Minuteman silos, 100 B-1 bombers, production of the D-5 SLBM, development of the small road-mobile ICBM, and improved surveillance, command, control and communications to ensure the validity of the SLUCA doctrine. It all sounded familiar.

The HASC strategic force structure improvements had evolved directly from my I/O–3 Study, my HASC testimony in 1982, and consultations with their staff members. Since they could not predict the outcome of the PCSF deliberations or the acceptance of the Commission's findings by the President, DOD and the Joint Chiefs of Staff, they wanted at least one supportive witness for the HASC positions. The first time I had given testimony to the HASC it took a year of petitioning. Being firmly asked to testify again was a welcome change.

The Scowcroft Commission had been afforded two unique offers. While the Commission might have come to similar conclusions without these inputs, we likely saved them a lot of time and discussion on a project that was subject to intense public interest and various organizational pressures.

The Commission had been independently provided a comprehensive paper with logical, defensible and specific recommendations. Coincidentally, the HASC then provided them with assurances of support for all of these suggested force structure improvements. The combination was hard for the Commission to resist. They essentially concurred with our conclusions, not knowing that the same analyst had been influential in both the paper and the evolution of HASC positions.

To Harden or Not to Harden

On February 16, after the HASC meeting with the Commission, Jim Woolsey advised me, with appropriate circumspection, that the paper was under much discussion within the Commission and that they were generally moving in that direction. This included development of the mobile ICBM that was not part of President Reagan's initial plan in October 1981. However, they intended to let the Air Force harden the missile transporter to facilitate basing in large, established garrison areas (a total of about 33,000 square miles) to minimize public interface with the mobile weapon system.

Intrinsically unnecessary, hardening of the missile transporter would substantially increase the cost and time required for the development, production and deployment of the missile. The Air Force wanted garrison deployment to large base areas and less deployment via the public highways. This would delay the small mobile ICBM program and thereby reduce the threat to the more important MX deployment and competition for MX production funds. Under the Air Force concept, the small mobile missile, now called the "Midgetman," would cost more than twice as much and take twice as long to deploy than the program described to the HASC in 1982.

I wrote a new Secret paper for the Commission showing that hardening the Midgetman transporter was unnecessary. It is doubtful that the second paper was distributed to all of the Commissioners; probably it was read by only Jim Woolsey, Marv Atkins and maybe General Scowcroft. Woolsey soon advised that, to a technologist, the paper was probably correct. However, the mobile missile public interface problem could not be ignored by the Commission. They would endorse the 20 psi hardening of the transporter for that reason. In their view, the added time and expense of hardening the Midgetman transporter for garrison deployment was a political necessity to ensure public acceptance of the weapon system deployment.

In his usual courteous style, Woolsey advised backing off on this and other issues since the Commission would eventually agree with about 95 percent of the original recommendations. He also said this advice applied to Colonel Pence. Larry had apparently also been at work in these matters independently. We would have to just wait and see.

Knowing that the Commission's findings would be subject to Presidential, DOD and congressional approvals, I delivered the second PCSF paper to the same people in these agencies who had received the first PCSF Paper in January–March. They were also advised that the overall progress and direction of the Scowcroft Commission's work was thoroughly satisfactory and that the paper on Midgetman hardening should only be useful if their total package became too expensive.

Commission's Report

General Scowcroft's letter to the President on April 6 stated that after 28 full meetings and many smaller conferences, after talking to over 200 technical experts and consulting closely with members of the Congress, they had reached unanimous conclusions and recommendations.

On April 11, Ron Mann called from the White House and said that I could pick up a copy of the Commission's report. The Commission had approved of everything important to the immediate and ten-year improvement of our strategic military posture!

I discussed the report with Larry Pence and others in preparation for my subsequent testimony to the HASC. The *Report of the President's Commission on Strategic Forces*, April 1983, stressed and resolved the issue of MX basing in the Minuteman silos and approved development of the hardened small, road-mobile ICBM. They also approved the rest of the President's new force structure, without the same detailed analysis afforded the ICBM force.

On April 19, President Reagan announced his decision to support the Commission's findings. During subsequent HASC hearings on April 20 and 21, Secretary of Defense Weinberger and the Joint Chiefs of Staff also stated their acceptance of the Commission's positions regarding all strategic forces. The five-year cost of the approved programs would be $19.9 billion in FY1982 dollars.[1]

House III

I set to work re-verifying intelligence data and preparing my own statements, in particular the addendum noted the futility of Closely Spaced Basing (CSB) and the unneeded hardening of the Midgetman Transporter, if necessary to conserve funds. I also emphasized the urgency of implementing certain of the Commission's findings to regain strategic parity through force improvements and arms control negotiations with the Soviets.

I sent the paper for the President's Commission to Bob Evans at IBM Corporate Headquarters and Chuck McKittrick, still the IBM vice president for government programs in Washington, DC. Three copies of the testimony paper were given to Ron Mann at the White House on April 13. My disagreements with the Commission's report were noted regarding deployment of the small ICBM five years earlier without hardening the transporter.

The final HASC hearings on the President's Commission on Strategic Forces would be April 20–22. Only four parties were to testify: the Honorable Brent Scowcroft for the Commission; the Honorable Casper Weinberger, Secretary of Defense; General John W. Vessey, Jr., Chairman of the Joint Chiefs (accompanied by the chiefs of the Army, Marines and Air Force); the Chief of Naval Operations; and finally, the untitled L. C. Carpenter.

Now, I owed an explanation to Jim Woolsey. Actually, he had noticed that the HASC's recommendations were practically identical to the paper I had prepared at his request and assumed my involvement with the HASC staff prior to their briefing to the Commission. I also told Woolsey of my commitment to the HASC to testify in support of their position — whatever the final positions of the Commission, President Reagan, Secretary of Defense Weinberger, and the Joint Chiefs might be. Jim said that General Scowcroft did seem a little surprised at the probable connection.

Hearings

Chairman Melvin Price opened the HASC hearings as scheduled on April 20. His statement was followed by that of Congressman William L. Dickinson (R-AL), minority member. The proceedings are recorded in GPO Document No. 21-6820, Hearings on H.R. 2287, DOD Authorization of Appropriations for FY1984; Committee on Armed Services, House of Representatives, 98th Congress, First Session, Part 2 of 8,

Strategic Programs. Currently unemployed, I attended all of the hearings in preparation for later testimony, reported on p. 181–221.

Lt. General Brent Scowcroft was accompanied by Jim Woolsey, John Deutch and John Lyons from the Commission. The General's presentation of the Commission's report and subsequent Q & A were well-received and very satisfactory to the HASC (p. 31–103).

Secretary of Defense Weinberger's statement and acceptance of the Commission's findings were reported on p. 105–158. After two years of his ineffectual leadership that led to the President's Commission and these hearings, he had quite a lot of explaining to do. However, there was no acrimony regarding his performance.

General John Vessey, US Army, chairman of the joint chiefs of staff, presented the unanimous Joint Chiefs' acceptance and rationale for their support of the Commission's report. He was accompanied by General Edward Meyer, chief of staff of the Army; General Robert Barrow, commandant of the Marine Corps; General Charles Gabriel, chief of staff for the Air Force; and Admiral James Watkins, chief of naval operations.

The Soviets

After my testimony to Senator Laxalt's subcommittee was published in 1981, wherein I had asserted the apparent malevolence of the Soviet Union, a call came from a friend in the intelligence community. He advised that this testimony had established me in the status of an "enemy of the Soviet State." He cautioned me to look both ways when crossing the street and never to go into the USSR on a civilian (blue) passport. My response was due caution, for a time; but I had never had any desire to visit the Soviet Union under any circumstances. Herman Kahn later provided the double translation paper that confirmed the Soviets' continuing interest in my Senate testimony of November 1981.

While I listened to the Joint Chiefs' testimony on Thursday afternoon, April 21, two strangers sat next to me. The spokesman, who identified himself as "Val" from the Soviet Embassy, was quite familiar with my prior testimony given to the House and Senate Armed Services Committees during 1981–1982. The KGB obviously had my picture and the prior testimony in their files. He wanted to discuss that testimony and extended a luncheon invitation at the Soviet Embassy, which I declined. Val said that these testimonies could only be given by an agent of the CIA. I asked him to either be quiet or leave. They left.

Testimony

Testifying to the HASC, again in Room 2118 in the Rayburn House Office Building, I gave an abridged summary of the paper to the President's Commission, emphasizing my March 1, 1982, testimony to the HASC and the subsequent gradual acceptance of the I/O–3 Study findings and recommendations. These included MX basing in the Minuteman silos and the small mobile ICBM. Hardening of the mobile transporter was noted as intrinsically unnecessary except to facilitate garrison basing of the Midgetman.

I answered questions from Congressmen Price, Stratton, Badham, Ray, Hopkins and Stump and from Staff Director John Ford and Alan Chase. All of their questions were designed to probe matters that they particularly wanted emphasized in the hearings record.

Congressman Stratton (D-NY) asked for a comment on the "build down" concept being pushed by Congressman Al Gore (ever the policy wonk) to replace MIRVed ICBMs with single warheads. I noted that the cost of such an arms agreement, while maintaining a comparable degree of mutual deterrence, would be prohibitive to both the Soviet Union and the United States.

This question had been anticipated. I quoted from a statement by the head of the Soviet Academy of Science's USA Institute, Georgi Arbatov, April 9: "The Reagan Administration is trying to provoke the Soviet Union into doing something harmful to the Soviet economy, negate our capital investments, and make us spend so much that we will bleed white." I added, "If the Soviets want to redeploy the forces they have today in single warheads ... they would have to have 6,000 hardened single silos, which is utterly illogical and prohibitively expensive. They already have everything that they need."

This occasion was the culmination and near-epiphany of my 30 years as a voluntary consultant to the US government.

B–1 Bombers

After two decades of support to the B-1 Program, I had never seen the aircraft. Major General Douglas P. Nelson (Ret.) remembered my long and finally successful B-1 advocacy efforts; he had initially sponsored them as B-1 SPO Chief in the early 1960s. He endorsed a professional or "VIP" visit to Dyess Air Force Base in October 1997 to include witnessing two B-1B training flights. I spent an hour in and all around a B-1B. The B-1B weighs over 250 tons with a full fuel and combat load, including over 33 tons of nuclear-armed missiles and gravity bombs. The four engines have so much power that they can propel the plane from brake release to 600 MPH in about two minutes.

Forty-four B-1B bombers faced each other and were parked wing-to-wing on the tarmac of the main taxiing runway. The two rows of bombers were both more than a mile in length. Each B-1B has three separate bomb bays containing eight-weapon rotary launchers for missiles or gravity bombs. These 44 bombers, alone, are capable of delivering more than 1,000 nuclear weapons to any potential enemy in Eurasia.

Take-offs at night are spectacular since the multiple diamond-shapes in the engine afterburner plumes extend rearward for about 200 feet, longer than the length of the aircraft. During my visit, the pilot decided to demonstrate a "short-field takeoff" such as might be used if they were under attack. This meant taking off downwind, using only half of the runway! When an aircraft has that kind of thrust, you can make your own rules. It also meant that we in the "viewers' vehicle" were now inadvertently parked in the second B-1's taxiing path. This monster of an airplane, towering four stories above us, headed directly toward our car. Given the B-1 pilot was only a major, and our driver was the AFB Vice Commander, Colonel Heiser I pointed out that we had the right of way. Even so, we quickly exited the taxi-way beneath the B-1's left wing.

The colonel knew what was coming next and drove us back to our previous viewing area so we would be able to take pictures outside of the car. The B-1 passed us, flying at about 200 feet on full afterburners. There was a brief blast of heat and our internal organs seemed to vibrate from the intense and deafening engine noise. This was an extremely satisfying, gut-rattling end to the B-1 tour.

Half of the B-1 bomber force is located at Dyess AFB to ensure maximum tactical warning of an ICBM or SLBM attack. They would have less than 30 minutes for all of the B-1s to leave the airbase if under ICBM attack and less than 20 minutes if depressed-trajectory SLBMs were launched from submarines in the western Atlantic or eastern Pacific oceans.

A large number of C-130 transport aircraft are also located at Dyess AFB, apparently to provide logistical support to a dispersed fleet of B-1s. On strategic warning (a few days, not minutes), both the C-130s and B-1s would be dispersed to alternate commercial and military airfields. The immense thrust of the four B-1 engines in short-field takeoffs would allow them to use any of some 600 such airports, thus precluding effective pre-emptive or trans-attack targeting by any enemy.

The trip to Dyess AFB was indeed full repayment for my 21 years of B-1 advocacy.

Chapter 20. US Arms Control and Disarmament Agency, ACDA

Now began my real work for the ACDA, on the fourth floor of the Department of State Building. Within three weeks of being hired, I had nine separate projects, old ones I took over and new ones that began with my arrival. Back to working long hours and weekends — the many severe stress symptoms of the last years in IBM had all returned.

Ambassador Adelman had recently testified to Congress. He returned to ACDA with 25 questions from the senators and congressmen. I was the Action Officer on this three-month task.

Verification is always a major issue in arms control that required several on-going studies. I was also assigned to ACDA's overall verification efforts and soon as a helper for Vigdor Teplitz, the senior bureaucrat responsible for a long-running study of anti-satellite weapons and policy. This involved the ACDA Working Group (WG) and later on the Inter-Agency Working Group (IG). The latter included all of the major government agencies concerned: ACDA, State Department, DOD, NASA and the CIA.

The thorough access to classified information in all areas of my new responsibilities included periodic intelligence briefings at the Pentagon and State Department. Data was easier to get — the handling of classified documents within the government was generally less restrictive than in industry. However, Sensitive Position employees were required to have "code word" clearances beyond Top Secret, for access to intelligence data from surveillance satellites, the Atomic Energy Commission, and NATO. The applications and approvals required five months.

Vic Alessi asked me to provide more help to the Deputy for Verification, Al Leiberman (VI), in establishing new classified and unclassified databases. I solved this problem at the Pentagon by obtaining a history of all US military forces since the Revolutionary War. The classified portion of this huge loose-leaf notebook contained a list of all US nuclear weapons produced through 1982 and those currently operational or in storage.

As a member of the ACDA Bureau of Strategic Programs, I was expected to provide contributions to all aspects of strategic arms control for missiles, aircraft and space weapon systems. By mid-April, I was writing a paper for Lou Nosenzo summarizing the ACDA's position on Anti-satellite (ASAT) weapons systems, working with Vigdor Teplitz's Inter-Agency Group (IG) on ASAT, and negotiating our differences with the CIA.

ASAT–Option 9

Under Teplitz, the study of anti-satellite (ASAT) weapons systems had expanded into a vast and comprehensive paper that was essentially useless in the formulation of policy. Vic Alessi asked me to get to the essence of it. I studied Vigdor's eight options to find a common thread; then I condensed them into something more actionable.

On Saturday morning, June 16, the *Washington Post*'s front page carried an interview with a senior State Department official regarding future ASAT arms control policy. The interview summarized the work statement I had drafted for Option 9 a few days earlier — the leak was not my doing.

Verne Wattawa

After several months, I moved to an office shared with several other ACDA senior professionals, next door to Lt. Colonel Verne V. Wattawa's office. It was eight years since I had last seen him, as a major working for Major General George Keegan in the Intelligence Air Staff.

In the interim, he had commanded a 275-person intelligence group in the Republic of Korea. His mission was to plan and guide the collection, analysis and production of US Air Force intelligence and develop targeting plans for the defense of Korea and the Western Pacific Region. In this position he periodically gave intelligence briefings to ROK President Rhee.

When he was assigned to ACDA as an Air Force intelligence officer, Verne later retired to civilian (GS-15) status in much the same position. We worked together on several assignments, particularly the means to monitor and verify small, mobile ICBM forces.

By now I also shared the responsibility to be the weekend Duty Officer for the Agency. This consisted of occupying the Director's office on the fifth floor of the State Department (first occupied by the Secretary of State, it is a vast room rich in marble décor, antique furniture, and high ceilings) and reviewing the three-inch stack of worldwide government cable traffic. If the Duty Officer detected any events that warranted an immediate ACDA response, he was to call Ambassador Adelman or Deputy Director Dave Emery.

SDI Tour

On September 26, Hank Cooper sent me on a tour of the facilities supporting President Reagan's Strategic Defense Initiative (SDI) with the START delegation. The delegation was led by Ambassador Edward Rowney (Lt. General, Ret.) and Ambassador Sol Polanski.

About 30 people met at Andrews AFB on October 8 for a C-130 flight to Los Angeles, from which we visited the nascent Space Command site near Colorado Springs and the NORAD Headquarters in Cheyenne Mountain. This latter cavernous facility contains offices and living quarters mounted on huge metal springs to absorb the shock of a nuclear attack. There was a space shuttle in orbit during our visit. We watched shuttle's progress as it transited the Earth on the huge display that would be used for monitoring any nuclear weapons being delivered against North America.

On October 10, we visited the Lawrence Livermore Nuclear Laboratory in Livermore, CA, where Dr. Edward Teller initiated the work on the hydrogen bomb. We were hosted by Mike May, one of the two inquisitors during my testimony to the Townes Committee in 1981.

As part of the LLNL analytical efforts on nuclear war (and showy briefings to more impressionable visitors), Mike had developed a computer-driven model and large display that progressively showed a hypothetical Soviet ICBM attack on the US. SAC and NORAD also had such trans-attack displays, but theirs were much better battle-management tools. I offered Mike my mathematical model for that event which was more precise (but less dramatic).

We next flew to Albuquerque, NM, and the Sandia Nuclear Laboratories. We toured a then-secret facility doing research on the generation of thermonuclear power. The building contained an immense radial array of high-energy lasers that were all focused on a pea-sized pellet of nuclear material. These pulse lasers were simultaneously discharged into the pellet to generate an extremely large burst of thermonuclear plasma of a few microseconds duration.

The last day of our tour was spent at the High-Energy Laser (HEL) facility at the White Sands Missile Test Range in New Mexico. As an extension of their high-powered laser research at the Naval Research Laboratory, the US Navy had built a very large HEL. The test installation had a potential use as an anti-satellite (ASAT) weapon system.

Negotiating with the Soviets

In their January 1985 meeting, Secretary Shultz and Foreign Minister Gromyko agreed to resume arms reduction negotiations in Geneva, Switzerland, in mid-March. This would end 15 months in the doldrums and give more urgency to our preparations.

I began to focus again with Vigdor on ASAT. Prospects for new Geneva negotiations brought me six new tasks, including one on the Soviet aerospace defensive systems.

The US Nuclear and Space Talks (NST) delegation consisted of three major components: Strategic Offensive Forces, Defense and Space, and Intermediate Nuclear Forces (INF). Ambassador Max M. Kampelman headed the US delegation and Ambassador Karpov was usually the senior Soviet representative. Ambassador Hank Cooper was responsible for our D&S delegation.

Verne went to Round I as the Advisor for D&S and in more important roles for most of the other rounds. For the first two rounds of the negotiations I provided some support via the secure telephones and prepared several position papers needed for subsequent rounds.

Alessi assigned me to back-stop the D&S delegation during Round II from May 31 to mid-July. This meant gathering up the overnight cable traffic and diplomatic pouch at 7:00 a.m. and preparing summaries for Lou Nosenzo's staff meeting at 8:15 and Alessi's at 8:30. Both meetings scheduled action items in immediate support of the delegation.

Supporting Groups

The first eight months of 1985 included over 100 WG and IG meetings at ACDA, State and the Pentagon. With negotiations underway in Geneva, these meetings were more frequent and intense than during 1984. Participation now included ten separate government agencies, such as the Office of Management and Budget (OMB) and the President's National Security Council (NSC).

Since both the Departments of State and Defense were vitally interested in the direction and scope of the Geneva negotiations, the Strategic Arms Control Policy Interagency Group became very involved.

Japanese Ambassador Imai arrived at the East Entrance to the State Department on June 27 to discuss US arms control policy with ACDA Deputy Director Dave Emery. As the Action Officer and note-taker for his visit, I asked the guard whether I ought to greet Ambassador Imai outside or within the reception area. The guard's pragmatic reply was, "Greet him inside if you think someone is going to shoot him."

NORAD & DSB

On another visit to the Cheyenne Mountain Headquarters of the US Air Force Space Command, the commander, General Herres, and his staff gave us several useful briefings on US and Soviet space activities. Our group also visited the growing new facilities a dozen miles east of Colorado Springs. Sprawling Peterson AFB was to be the new headquarters of the Space Command. Fifteen years after the Cold War ended, all but the emergency functions of NORAD were also moved to the new AFB.

The Defense Science Board 1985 Summer Study session met, as usual, at the Naval Ocean System Center atop Pt. Loma in San Diego. I represented ACDA at one of the panels considering Tactical Directed Energy Weapons, which also had the potential for ASAT applications and possible countermeasures, during a day of meetings on high energy lasers (HELs). Vince Cook also attended for IBM, but we had nothing to discuss.

Public Diplomacy

After several weeks of training, I became one of 150 government officials authorized to publicly speak on arms control matters. I gave TV and radio interviews and formal speeches in Idaho, North Carolina and Virginia.

Giving the Reagan Administration view (which I shared) of the status of relations, I would observe that the START Talks were in limbo because the Soviets had deployed 378 of their SS-20 mobile IRBMs, each armed with three nuclear warhead MIRVs. Of these missiles, 243 threatened NATO and 135 more were deployed against our Asian allies and US military installations in the Far East. In December 1979, NATO requested US deployment of 108 mobile Pershing II IRBMs and 464 Ground-Launched Cruise Missiles (GLCMs) to offset this latest Soviet threat. We later initiated the 1982–1983 talks for reductions of these Soviet and NATO missiles. The Soviets discontinued these talks on November 23, 1983, after their diplomatic and public campaigns failed to stop the Pershing II deployments. On December 8, 1983, they also refused to reconvene the START negotiations.

The Soviets did not respond to President Reagan's speech to the United Nations General Assembly, on January 16, 1984, until June 26. They called for talks in Geneva "to prevent the militarization of outer space." The wrangling continued without negotiations until a meeting of Secretary of State George P. Shultz and Soviet Foreign Minister Andrei A. Gromyko on January 7–8, 1985.

I would describe the Soviets' unilateral 15-year effort that led to their strategic military superiority. Then I talked about our new strategic forces modernization programs, SDI, and possible negotiations to re-establish parity, and dismissed the notion of "nuclear freeze."

Our policy derived from the leading political and military thinkers in the US during that era and was credited with providing the basis for later victory in the Cold War. In a nutshell,

> The US approach to recent and anticipated arms control negotiations is based on three guiding principles: realism, strength and dialogue. You have been offered realism on the nature of our adversary. Our strategic modernization program will renew the strength of our nuclear forces to ensure that we can deter major Soviet aggression for the rest of this century.
>
> Our eventual success in achieving balanced and verifiable arms control agreements that make a real contribution to global stability and security require Western patience, persistence and unity. The US and our allies have led in these endeavors for more than 30 years. Pursuit of a more stable peace through a vigorous arms reduction program will remain among the highest priorities of the US Government. We solicit your understanding, patience and support.

Chapter 21. NST Round III — Geneva

Negotiations were scheduled for September 15 to early November, 1985. A few of the professionally ambitious people had attended Rounds I or II to later appear in the media as experts on arms control. The negotiations were now past this phase and were entering the years of grinding, slow progress; Ambassador Cooper called and told me to get a diplomatic (black) passport and security clearance transfers.

Verne Wattawa and I started work on D&S plenary statements for Round III and cleared these papers with the IG on September 12. We went to Andrews AFB before dawn and were on our way in a huge C-141 Air Force transport plane. All passengers were provided with blankets for the cold, barn-like cabin and earplugs for the deafening turboprop engine noise. This was not luxury travel. The huge cabin contained a large, movable cube structure that was heated for VIPs, ambassadors and military general officers. As a GS-15, my rank for protocol purposes was equivalent to that of a colonel (0-6 Level). When the rank-heavy delegation was in session, the 0-6 Level was a qualification to brew the morning coffee.

At Geneva, a bus took the weary delegation to the Botanique Building, our temporary office near the city botanical gardens. A new US Mission building was being completed on a hill west of the United Nations, International Red Cross and USSR Mission complexes. The first working day started in the "bubble" at the Botanique Building. The bubbles were rooms within rooms specially constructed to thwart sophisticated technical eavesdropping.

Ambassador Kampelman

Max M. Kampelman, chief negotiator for our overall delegation, attended some of the sessions. He asked me to visit his office after the working meetings were completed. The Ambassador knew more about the I/O–3 Study than I had expected; he said that there would be special tasks to be performed, only for him, and that the work should not be discussed with anyone, including Ambassador Cooper. This was uncomfortable, but "Max" (only peers called him Max in direct conversation) had good reasons for such an arrangement.

Ambassador Kampelman needed prompt and dependable analyses of new situations without the doctrinaire response and delays of the delegation and Washington. My work was to cover all three major aspects of the negotiations: strategic offensive and defensive forces, and INF.

The delegation make-up was about half military people and the rest were civilian bureaucrats and appointees. The civilians included many very conservative political appointees, such as Ambassador Cooper. Consequently, the majority of the US delegation was understandably wary of any but the most favorable terms in our negotiations with the Soviets. Some of the military participants were direct and deliberate obstructionists to any arms control initiatives.

The more constructive agencies in the US delegation included the State Department, headed by their member, Greg Suchan, and ACDA as represented by Hank Cooper, Verne Wattawa, and others. Bob Barry ably headed the intelligence community for the entire delegation and by definition or law was a neutral and constructive participant. Bob had had much the same job during the SALT negotiations in the late 1970s. Although *Time Magazine* blew his cover then, he and other CIA people were listed as State Department employees.

First Plenaries

The negotiations had been thoroughly structured and formalized before and during Rounds I & II. The most important plenary meetings were chaired by Ambassadors Kampelman and Karpov, Obukhov or Vitsinsky for the Soviets. The ambassadors both read formal statements and led the subsequent discussions. Kampelman's delegation consisted of an ambassador for each group, e.g., Ambassador Cooper for the D&S Group, and a dozen or more 0-6 Level military and civilians. The 0-6s sat behind their respective ambassadors and did not speak to the Soviets during the plenary. The higher level SES employees and one- or two-star generals sat at the table in support of their ambassadors. Military personnel on both sides wore civilian clothes during the negotiations.

The protocol of negotiations required that each nation be referred to as a "side," e.g., the Soviet Side. This practice allowed a distinction between delegation discussions and the formal final position of each nation on an issue. US–Soviet meetings alternated between the Soviet Embassy and our Botanique Building offices.

The D&S delegations held the first Round III Plenary Session on September 18; the first Joint Plenary Session for Round III was held the next day. Much of the prior two days (and nights) were spent communicating on the secure telephones to Lou Nosenzo and Vigdor Teplitz at ACDA. As usual, the NSC had not fully approved of the delegation's instructions for Round III. We used "back-channel" secure military communications for transmitting written messages to confirm our telephone calls. Back channels were both secure and more private than the secret, formal daily cable traffic that required collective approvals.

Obtaining final, timely White House (NSC) approvals of each delegation's instructions at the beginning of every round was always difficult. Interim work by the IGs was often contentious and their agreed written positions were likely supplemented by private calls to friends on the NSC. The Soviets appeared to have similar problems. Each side occasionally had to hold their first plenary session without any new instructions. This was both obvious and an embarrassment.

The daily process in the delegations was much the same as that of working groups in Washington. However, the meetings were much more intense due to deadlines imposed by scheduled meetings with the Soviets. There was a separate floor for each delegation with representatives from ACDA, State, DOD, Joint Chiefs of Staff (JCS), and the intelligence community on each floor.

After, or without, D&S Plenary Sessions, less formal negotiations went on between US and Soviet working groups. These were the best means to really get to know your opposite numbers. The two Russians I normally encountered were Vladimir Grinin and Aleksandr Klapovskiy. Grinin was a Soviet "lawyer" with close ties to Ambassador Kvitsinsky. In later rounds, the Soviets blew Klapovskiy's cover as a colonel; probably in the GRU, the internal Soviet security organization. On September 24, I was seated at the table with Captain Ed Melanson (USN) and Paul Lembesis, an ACDA expert on international law. Following the plenary, this time at the Soviet Mission, we had the usual working group meeting; I made the customary notes for our Memorandum of Conversation (Memcom) and reporting cable to Washington. Pete Zimmerman, our back-stopper at ACDA, was also called and given an oral report on the meeting. The memcon was critiqued by Colonel Bob Moser, a long-time DOD arms negotiator and several other people on the delegation; the cable finally cleared by 7:30 p.m.

All of these US–Soviet meetings were augmented by "social" occasions — lunches and dinners in Geneva or outlying restaurants between small groups with interpreters from both sides. In diplomatic circles, there is no such thing as a simple social gathering. You are always "on," looking for nuances in their positions or very minor "breakthroughs" that could be realized in these more relaxed and alcoholic gatherings.

I completed my first study for Ambassador Kampelman on October 6. Next, Max asked Colonel Moser to provide me with an updated Air Force summary of current and planned strategic weapon systems. Noting that Colonel Moser's data was more current than that provided by the CIA, I updated my October 6 study and gave a copy to Bob Barry. Bob, by definition and trust, was part of Kampelman's inner circle.

The two papers were generally concerned with "arms reductions and Soviet initiatives" and possible US responses. On October 15, Ambassador Kampelman sent a memo to Hank about the use of the two papers. Ten minutes later, Hank was in my office asking for his copies. Hank probably was not pleased to learn of my private work for Max, but he was too much the gentleman to discuss it.

I met separately on October 23 with Max and later with Hank. While the papers could not be furnished to or used directly by the delegation, the viewpoints expressed had some influence on our work through the directions provided by Max and Hank.

Due to the Swiss' famous and enduring neutrality during the two World Wars, Geneva was believed to be the Central European Headquarters of the Soviet KGB. There were estimated to be several hundred KGB and Warsaw Pact agents in this international meeting place of about 200,000 people.

The local wags noted that for every diplomat there were two reporters and three spies. There was little actual danger from all of these professional spooks. Terrorists were another matter. The local mood gradually turned for the worse during the next several years due to terrorist activity throughout Europe, particularly in 1986.

We avoided discussing classified matters in our cars. Mine was parked in the hotel's basement garage, next to that of a known colonel in the KGB. Listening devices in cars are very difficult and expensive to detect, so automobiles were seldom "swept" except for high-ranking officials. We always assumed that our cars were "bugged."

Having attended the State Department's anti-terrorist courses, we varied our commuting routes to the US Mission along with other customary precautions.

Most of us did feel a little on edge; but despite having gotten a number of death threats Ambassador Kampelman persisted in taking his preferred route across the lake by water taxi and walking more than a mile up the hill to the US Mission for private time to think and get his daily exercise. After the assassination of Kampelman's friend, Prime Minister Palma of Sweden, and with increasing threats, Max did change his ways. Other senior members of the delegation, including Hank Cooper, also began to use their very uncomfortable three-ton armored Chevrolets more frequently.

Little real progress was made by the fifth plenary session in late October. I began to make modest contributions to the plenary statements. My two Soviet interlocutors were also more frequently engaged during our working group sessions.

Senator John Tower, now Ambassador Tower, was head of the START Negotiating Group. At our first meeting in the Botanique Building, I made the obligatory self-introduction to Tower, acknowleged by a contemptuous grunt. We had never met before, but he was well aware of my several testimonies to Congress during 1981–1983. My work with the Congress had defeated the positions he preferred when he was chairing the Senate Armed Services Committee. He hadn't forgotten.

Soviet Surprise

As in the US, I was occasionally assigned as the weekend Duty Officer for the overall delegation in Geneva. The routine was the same: collect and read all of the cables and any diplomatic pouches. Further, if there was anything that required a prompt response, I was to call the ambassador similarly designated for weekend duty.

On Saturday, October 26, during my third tour as Duty Officer, there was a historic jolt in the cable traffic. The Soviets were offering a joint ten percent reduction in ICBM warheads! This was the first real initiative by either side during seven months of negotiations. The timing, on a weekend late in Round III, was intended to evoke our prompt attention. I called Ambassador Tower immediately.

When Tower arrived at the Mission, I assisted him in sending notifications to the State Department, ACDA, and the other local ambassadors and senior professionals in the delegation. After the initial flurry of activity was over and Senator Tower left, I began to make a comprehensive analysis of the Soviet proposal and possible US responses. Ambassador Kampelman would not want to wait four to six weeks for the State Department, Joint Chiefs of Staff, and the NSC's final position. As Deputy Negotiator for D&S, Hank was involved, but he was not immediately responsible for generating a position on the matter.

On Monday, I met privately with Ambassador Kampelman to discuss the approach I was taking on the analysis of the Soviet initiative and received his encouragement. In the wee hours of Thursday morning, I was almost done when I had another attack of a severe headache and scintillating scaratoma. I couldn't see well enough to continue; I placed the text in Valerie Daniel's safe with a note asking her to type the nearly completed papers.

After the day's negotiations, I discussed the papers with Ambassador Kampelman. The position offered was essentially to negotiate with the Soviets to reduce a spectrum of their ICBM warheads, primarily on their counterforce-capable SS-17s, SS-18s and SS-19s. In response, the US would eliminate the same percentage, but a lesser number, of comparable ICBMs. Three or more alternative mutual force reductions of this nature were included.

Ambassador Kampelman said he was inclined to agree but that this approach would probably engender a severe reaction from the delegation, ACDA, DOD, and Defense Secretary Weinberger. He also knew that cable traffic on the analysis was out of the question. Further, a back-channel message to the State Department through the military communications system on such a sensitive matter would likely be leaked to DOD.

After we pondered the situation, I suggested that on the pretense of being ill I should return to Washington and personally deliver the papers to whoever should have them. This would be quite believable since I had many chronic illnesses. Early the next morning his private secretary called me at the hotel and said that Max had directed the ACDA Administrator in Geneva, John Grassley, to provide airline tickets on the first available flight to Paris and then Washington, and to include a diplomatic courier pass. Bob Barry provided an illness cover story to the Delegation.

On arrival, the first stop was the State Department which had recently experienced a shooting incident; security was at a higher level than usual. However, remembering me from all my working weekends at ACDA, the guard readily waved me through.

On the Eighth Floor, in an office near that of the Secretary of State, I met President Reagan's Ambassador-at-Large for Arms Control, the venerable Ambassador Paul Nitze. In a half-hour meeting, we covered the essentials of the Soviet proposal for strategic missile warhead reductions and the several possible and positive responses.

Ambassador Nitze, while appreciating the effort and Ambassador Kampelman's endorsement, said that the Administration would not likely accept the analysis and arguments. As a 50-year senior diplomat, Nitze knew that the time required for protracted negotiations should not be a factor in such a decision. Patience must be near infinite. Further, he was concerned that the Soviets would try to broaden their proposal to assault our SDI Program, in Ambassador Cooper's area of responsibility.

ACDA

The delegation returned from Round III on November 7. Ambassador Kampelman invited me to his office on November 13. From his higher-level contacts, he was more positive about the results that we had obtained through our recent errand. He said that there would be no immediate response, but the seeds had been planted.

Advisors on the delegation were normally rotated after every round. Only Members of the delegation, such as Greg Suchan and Bob Barry, and the Ambassadors returned for every round, but Max said that he would ask Director Adelman to designate me as a permanent Senior Advisor to the delegation — while still working for Hank, of course.

On November 14, the *Washington Post* ran the headline "Quick Cut in ICBMs Proposed by the Soviets." Vic Alessi did not know of my liaison from Kampelman to Nitze, but routinely asked if I was behind this major leak to the press. Vic had long known of my continuing friendship with Bill Beecher, who was still a very prominent Washington reporter. I did not know the source of the leak, but judging from the content, the source was much higher in the government.

The *Post* also carried two more major articles. "Gorbachev Insists SDI is a Leading Summit Issue" and "Moscow Warns Bonn on SDI." SDI understandably was of major concern to the Soviets. They still referred to our NST Geneva negotiations as "Meetings to Prevent the Militarization of Outer Space."

The US was usually eight to ten years ahead of the Soviet Union in most military technologies, but until SDI we had never before been organized internally and directed to provide an effective ABM defense. The US had now allocated the resources with presidential support. If SDI was successful, the USSR was afraid that a US ABM defense would offset the strategic military superiority they had invested so much to achieve. They were also over extended in other military spending, including their invasion of Afghanistan and lacked the resources to mount a comparable effort.

Assignments

After we all returned from Geneva, Vigdor Teplitz came by with a new assignment for me, which I refused. On December 3 Colonel Bob Moser advised me that I would indeed return to Geneva for Round IV. Any tasks I performed for ACDA in December were directly related to the Round IV negotiations.

I had suggested to Ambassador Kampelman that he might find my I/O–3 Study briefing to be a useful perspective on the Reagan Administration's strategic arms initiative; and I briefed him and Hank on December 12. Hank had heard some of the story at the Townes Committee in 1981, but the briefing to the Committee had been very selective to eliminate their bias toward the MPS basing plan for the MX ICBM.

Although the I/O–3 Study was completed five years earlier, the intelligence data and projections that I used to predict conditions through 1985 were still valid. Since the study had thoroughly influenced the Reagan Administration's program to renew our strategic military forces, the two ambassadors found this report still offered a useful overall perspective on extant US and USSR strategic military capabilities.

In January 1986, all of the ambassadors in the arms negotiations became direct employees of the State Department. This included Ambassador Tower for START, Ambassador Cooper for Defense and Space, and Ambassador Glitman for Intermediate Nuclear Forces (INF). This vacated Hank's office in ACDA, so a new civilian manager transferred from the Pentagon. He was Mike Mobbs, a long-time associate of Richard Perle and Frank Gaffney in OSD. During the State Department ceremony related to the new management structure, Frank curtly indicated that he remembered our encounters all too well. With Ambassador Cooper transferred, Ambassador Kampelman and ACDA Director Adelman's agreement to give me a permanent position on the delegation was very fortunate. Vic Alessi and Lou Nosenzo, were also supportive.

Due to some unfavorable publicity about the delegation, Colonel Bob Moser had been replaced by USAF Colonel Kramer as the administrative manager. His job was the day-to-day administration of the delegation's needs and assets while reporting to the Executive Secretary, Ambassador Warren Zimmerman. Most of the permanent military people on the delegation during my four years in Geneva seemed determined to reduce my effectiveness in any way they could. Colonel Kramer excluded me from some plenary sessions and denied me the use of a car.

Most of the NST delegation departed via USAF C-141 on January 10, 1986, for another long, cold, noisy overnight flight to Geneva. One suspects this was a way to make more efficient use of Air Transport Command training flights. The delegation was again a group of very intelligent, highly motivated people with many diverse personal and professional agendas. The considerable risk of internal divisiveness and acrimony became a new reality during the three rounds of the NST negotiations in

1986. Some members of the US delegation were there to negotiate and others were there to obstruct. Among a dozen others, primarily civilians, there was a sincere attempt to find common ground for mutually beneficial arms reduction agreements.

I sometimes had to challenge the positions of the delegation's generals and lesser military officers during our bubble-room meetings, and my increasingly apparent activity as Ambassador Kampelman's private strategic forces analyst undercut their intended and traditional responsibilities in that role. I was now understood to be ACDA's senior advisor for the delegation.

Chapter 22. NST Rounds IV–VI

Round IV

Guided by the latest delegation instructions, the plenary and joint plenary statements were drafted by the designated member or advisor and his working group. This was followed by three to ten drafts that were critiqued by peers, members and eventually, the ambassadors. Separate negotiating strategy sessions were also held when our formal instructions provided sufficient latitude or ambiguity.

After the morning staff meetings on the day of the plenary session, there would be a pre-plenary meeting, the formal meeting at the US and Soviet missions, bilateral working groups after the plenary statements, a plenary de-briefing at the US Mission, secure telephone reports back to the government agencies in the US, back-channel messages when needed, and preparation of plenary reports and memoranda of conversations (memcons) from all of the working groups. The formal reporting cables were usually cleared and sent to the US within 48 hours.

Given several joint plenaries and six or eight group plenaries by each of the three separate negotiating groups, the foregoing routine was repeated at least a dozen times in each round of negotiations. Separate luncheon or dinner meetings with various groups of Soviet negotiators were also reported by secure telephone, if urgent, and later memcons.

This slow, grinding process of negotiations that mostly resolved minor differences ("The Devil is in the details") necessarily went on for years. Continuing negotiations revealed that the Soviets would not agree to major START (offensive) arms reductions without a simultaneous, contingent package of agreements (linkage) to "prevent the militarization of outer space." This obvious Soviet objective required better liaison between the US START and D&S negotiating groups.

Knowing my background and interest in all weapons systems, both offensive and defensive, Ambassador Kampelman had me monitor the daily meetings of the START Negotiating Group and provide reports to him, Ambassador Cooper, and the D&S Negotiating Group. This would include providing the START plenary statements in advance of their meetings with the Soviets, the START pre-plenary sessions and post-plenary de-briefings. This would detect any moves by the Soviets to affect the interrelationship of potential START and D&S agreements.

I thought that Senator Tower might object, due to our past differences. Max said that he and Tower had agreed on a D&S representative at the START meetings in the bubble rooms, but also implied that this task should be performed without ruffling any START feathers, particularly those of Ambassador Tower.

Bob Einhorn, the senior State Department permanent member on the START, and I agreed that I would attend the early part of the START pre-plenary and post-plenary meetings to obtain the substantive portion of their negotiations. When internal negotiations (arguments) began in the meetings, I would unobtrusively depart. Although Ambassador Tower was keenly aware of the new presence in his bubble meetings, we never spoke to each other. To avoid the appearance of being a one-man spying operation, I usually asked another D&S advisor to go with me. The arrangement worked without friction for several rounds.

Contention

USAF Colonel Dan Gallington was the OSD member of the D&S negotiating group. Dan was a lawyer, dedicated to the interests of DOD and the Air Force. He had been selected by OSD to protect their interests in the D&S negotiations, by any legal means. Captain Ed Melanson, USN, was a captain of US Navy destroyers and a perfect complement to Colonel Gallington. Although Ed was Dan's advisor, his rank, intellect and connections within DOD and the Administration made him much more than that.

Due to the recent murder of a US Army major by the Soviet occupation forces in East Germany, Dan and Ed refused to attend any "social" meetings with the Russians. Both of these men were very likable as individuals. However, they were dependable adversaries in the D&S Group.

Our first two D&S plenaries described a US initiative for a "confidence-building measure," a negotiation move to effect better US–Soviet relations short of any actual arms reduction agreement. We proposed mutual inspection of laboratories suspected of developing programs for the militarization of outer space. The US proposal was understood from the outset to have little chance of success.

Colonel Gallington was the OSD member tasked to draft our first plenary statement on Open Laboratories. Many of us believed that Colonel Gallington had his own "instructions" from the Pentagon and that they were substantially different than those of the delegation. His actions were too singularly disruptive to be those of an ordinary 0-6 Level OSD Member of the delegation. The draft he came up with was unusable. Such a poor product from someone as smart as Dan could only be deliberate. After another unproductive meeting on January 17, Captain Melanson rewrote and cleared the paper for our first D&S Plenary of Round IV, held on January 21.

Gallington's actions were a direct challenge to Ambassador Cooper's responsibilities in the D&S negotiations. More severe arguments erupted between them on January 21 and 23. After a few more weeks of this conduct, Ambassadors Kampelman and Cooper acted to remove Dan from the delegation. Secretary of Defense Weinberger intervened, and Gallington continued his obstructive role through subtler and more subdued practices.

Ambassador Kampelman told me to attend and report on every working group meeting in which Colonel Gallington met with the Soviets. I found that working in the groups headed by Greg Suchan and Bob Barry had been more constructive.

Disaster in Space

The US space shuttle *Challenger* exploded soon after lift-off on January 28, 1986, killing the seven astronauts.

Three days later, the first US–Soviet reception for Round IV was held at the Soviet mission, an imposing and well-sited compound "liberated" from the Lithuanians after their occupation by the Soviets at the end of World War II. As was my habit, I selected a Soviet counterparty. On this occasion, it was Mr. Grinin. He was standing alone by a folded curtain at the window-wall of their main conference and banquet room. When I offered to introduce him to a visiting US congressman, he vaguely demurred. I presume he could not leave his station near one of the many microphones hidden in the room.

Grinin immediately launched into his own agenda. Not surprisingly, since the Soviet manned space vehicle was a very close copy of our space shuttle, his first question was, "What caused the *Challenger* explosion?" His second and third questions concerned US space research on a hypersonic trans-continental ramjet vehicle and another "black" program that was primarily conceptual at the time. I was creative in formulating my denials. (A diplomat is sometimes defined as an otherwise honorable man sent abroad to lie for his country.)

I spoke with two more of my peers on the Soviet delegation; their dialogue followed the exact pattern of the three questions asked by Grinin. After the meeting, I wrote a special memcon to Bob Barry warning of a possible effort by the KGB to penetrate the security of the Scramjet and black programs.

Soviet Initiative

In the first Joint Plenary of Round IV on January 16, the Soviets laid out a very comprehensive proposal to mutually reduce both offensive and defensive strategic armaments. If we had felt we could trust the USSR, the whole package might have been a good opportunity for the United States.

The Soviet proposal consisted of 22 inter-related steps to effect real arms reductions. I began a qualitative (and later quantitative) analysis immediately, including a schematic flow diagram of the 22 inter-related parts of the Soviet plenary statement and the many implied consequences, consistent with my now-established role as senior advisor and ACDA deputy to the D&S Negotiator.

Hank recommended some improvements, to be followed by wide distribution of the chart to the entire leadership of the delegation.

In the first meeting of Round IV with Ambassador Kampelman, on January 27, we privately discussed both the Secret and unclassified flow charts of the new Soviet proposal. I also provided a classified summary of a meeting in ACDA I had earlier hosted for Lt. General Odom, then commander of the National Security Agency (NSA), responsible for monitoring worldwide communications for the intelligence community.

The Senate Foreign Relations Committee Observer Group, locally referred to as Codel, frequently visited the delegation in Geneva. Due to the time difference from Washington, they usually came to the US Mission very early the next morning. I knew some of them from earlier briefings and testimony given on the Hill. In Round

IV, they arrived in time for the morning pot of coffee, about 6:30 a.m. Then and in later rounds I escorted them to the bubble and gave a briefing on the current status of the negotiations.

When Ambassador-at-Large Paul Nitze was in Geneva, and because they wished to get a first-hand appreciation of our negotiations, several senators visited us during February 8–12. They included Senators Warner, Stevens, Kennedy, Gore, Moynihan and Pell.

As the ranking minority member, Senator Pell (D-RI) was avoided by the right-wing people on the delegation. Seeing Senator Pell sitting alone at a US social function and wanting him to feel more welcome, I took the opportunity to have a long talk. I found his views on disarmament both naive and simplistic: "One nuclear weapon is too many." Conversely, conservative senators and congressmen tended to distrust the ACDA (and our efforts to negotiate arms reductions).

Ambassadors Cooper and Kampelman held several private parties for the delegation at their residences. These were always pleasant affairs without the stress of social occasions with the Soviets.

At Hank's request, I drafted the End-of-Round Report for D&S and summarize the plenaries. The drafts got lighter editing than usual — everyone wanted to go home.

Our C-141 military transport returned to Andrews AFB on March 7. Someone had prevailed on Ambassador Kampelman to fly with us. That was the last time the delegation flew only on C-141s.

First Interim

The time in the US between rounds in Geneva was used to prepare instructions for Round V, obtain new data on US strategic forces, initiate a new study for Ambassador Kampelman, and do some local tasks for ACDA.

Kent Stansberry, the OSD principal in the Defense and Space Inter-Agency Group, was preparing a paper on the Strategic Defense Initiative (SDI) that we needed for the Round V instructions. SDI and the Soviet ABM programs were increasingly contentious issues in the Geneva negotiations. The Soviets still intended that any agreement to reduce strategic offensive forces would require the linkage of interdependent limitations on SDI to "prevent the militarization of outer space."

This work soon involved the National Security Council (NSC), so we began to hold our D&S IG meetings in the New Executive Office Building (NEOB) across Pennsylvania Avenue from the White House. OSD finally gave enough ground to allow the usual last-minute preparation of the Round V instructions.

Washington Lunches

I still met with Bill Beecher from time to time. His perspectives on current government circumstances were always valuable. We would also get together before some of his many trips to Moscow. I also met regularly with Lt. Colonel Bob Reisinger, now retired from the Air Force and running his own successful defense consulting business. I mentioned that I was having difficulty in obtaining information on the evolving US strategic force structure. In Geneva, the current ranking Air Force general on the delegation had refused to provide such data, knowing the information would be used in my private studies for Ambassador Kampelman. Resourceful as ever, Bob arranged a meeting with several officers in his former organization at AFSC Headquarters at Andrews AFB. I promised them any significant results from subsequent studies and they provided data from all of the latest projections of US strategic forces during three visits to their offices.

Transition Study

The Soviets were more forthcoming in Round IV, but no one in the US military had provided possible US responses. We would need a quantitative analysis of mutual force reductions and qualitative study of possible instability at several transition stages during such reductions.

At the end of Round IV, Ambassador Kampelman asked me for a broad study of how the US and Soviets could gradually reduce strategic offensive forces for mutual advantages in national security. Bob Barry would provide current and projected Soviet forces for the study. With help from Bob Reisinger and AFSC Headquarters, I had collected all of the data needed for the study.

I reviewed the study plan and analysis of US forces with Ambassador Kampelman on April 29. He agreed with the approach and said to continue the study during Round V in Geneva. He also arranged a meeting with Ron Lehman, who had just replaced Ambassador Tower as head of the START Group of the delegation.

Round V

This would be the delegation's first full round in the new US Mission in Geneva. The new office was on the third floor, near the offices of Ambassadors Kampelman and Cooper and the new Executive Director, Avis Bohlen. She was the daughter of "Chip" Bohlen, President Franklin D. Roosevelt's Ambassador-at-Large and confidant. Her predecessor, Warren Zimmerman, had become Ambassador to Yugoslavia.

My immediate tasks included drafting the first D&S plenary statement and being the delegation's Duty Officer over the Memorial Day holiday.

Round V included all of the D&S negotiating and START liaison obligations accumulated in Round IV. In addition, I did some special studies on space-borne nuclear reactors for Hank and the new "transient study" of progressive disarmament steps and possible instabilities that might exist in the US–Soviet balance of strategic forces.

Intensified negotiations now required immediate START pre- and post-plenary briefings to both Hank and Max. Late in the round, they both asked me to read and highlight the START cable traffic for them. All this ensured that I was well informed, but very tired.

There were seven D&S and START plenaries and several Joint delegation plenaries. The progress of the Intermediate Nuclear Forces (INF) negotiations were also followed, but without direct involvement in their proceedings.

Colonel Gallington and Captain Melanson were still adversarial but more subdued due to the confrontations in Round IV.

During the Memorial Day weekend as Duty Officer, the Soviets cabled a new initiative regarding the Carter-era ABM Treaty. On Saturday morning Hank and the D&S Members were all notified. An immediate analysis of the Soviet initiative warranted notifying Ambassador Kampelman. He was then back in the States for a respite and high-level meetings in the State Department.

A summary was dictated to Sharon Martin, his secretary, on the secure telephone. When Ambassador Kampelman returned to Geneva on May 29, we discussed how the new Soviet ABM proposal might affect my "transition study" work done during the past two months.

On June 11, there was a Joint Plenary at the Soviet Mission that I did not attend again due to machinations by the Air Force administrator, still Colonel Kramer. The Soviets surprised us with a very important integrated proposal for START and D&S that included the extant ABM Treaty. Ambassador Kampelman stopped by the office on his return to relate their new proposal. Bob Barry provided a copy of his notes, which I needed for subsequent analysis of the implications.

After two months the transition study was well along and incorporated both of the Round V Soviet initiatives. After a special effort related to the B-52Gs, I got Max the first hard copy on June 16. After the sixth D&S Plenary, we met to discuss the phasing, analysis and stability assessments. He first noted that "ACDA won't agree." However, he endorsed the methodology. He said to give the study to Bob Barry, to be back-channeled to Doug George who was then heading the intelligence community group that provided people to the delegation. These professionals included personnel from the CIA, DIA, NSA and other government agencies.

After further reflection, on June 19, Max directed that the entire transition study also be given to Ambassador Ron Lehman.

Earlier, during Round IV, Bob Barry had asked me as the de facto member for ACDA to share a lunch that the ACDA and the CIA would hold for our counterparts on the Soviet delegation at a lake-shore restaurant a few miles north of Geneva. This occasion brought out the first, minor, and only new thinking by the Soviets in Round IV.

Breakthroughs and nuances in the negotiations usually occurred only at joint plenaries, meetings between ambassadors, or at luncheons between members of the opposing sides. Since Bob Barry was a known CIA member, his counterparts in the KGB et al. were our luncheon guests.

We had two more luncheons at Geneva restaurants with the Soviets during Round V. They always reciprocated at the Soviet Mission, presumably to preserve both funds and security. The best US interpreter was Peter Afanasenko; Peter also served as President Reagan's interpreter on State occasions.

My gray hair, three-piece suits, and association with Bob Barry probably convinced the Soviets I was a CIA affiliation or was being co-opted by them. That was fine — a little undeserved professional dignity never hurts. Ambassador Kampelman also reinforced my credentials with the Soviets and with our own delegation. When Colonel Kramer did not interfere, my usual seating was immediately behind Max at joint plenaries.

Summer Break

The interval between Rounds V and VI was busier than ever. With Gorbachev as USSR Premier and a forthcoming President Reagan, there was greater hope for progress in the arms control negotiations. I went ahead with the usual ACDA work including drafting the Round VI instructions for D&S and START; but my most useful work was on START for Ambassadors Kampelman, Cooper and Lehman.

The three ambassadors were in Moscow meetings with Ambassador-at-Large Paul Nitze on August 11. On August 13, Ambassador Kampelman described the Moscow meetings to me and we discussed how this would affect my studies for him on START. Then he sent me to discuss my analyses and possible responses to the Moscow meetings with Ambassador Lehman.

My work for Ambassador Kampelman in Geneva had included the spectrum of possible arms reductions, including unlikely radical reductions of greater than 50 percent. Ron Lehman asked for subsequent analyses on various options of 50 percent reductions or less. Hank Cooper asked me to start with 30 percent or less.

The current buzzword for strategic weapons was "Strategic Nuclear Delivery Vehicles" (SNDVs). Since this could be interpreted as missiles or guided nuclear warheads, I provided analyses that resulted in mutual percentage reductions of boosters, aircraft and maneuverable warhead delivery vehicles, whether MIRVs from ICBMs and SLBMs or ASMs from strategic bombers. This complicated the analysis but insured that the US would not accept disadvantageous percentage agreements. Any linkages to aerospace defense reductions were treated separately.

Instructions to the arms control ambassadors continued to evolve through the Inter-Agency (IG) process. Two high-level arms control meetings were pending, one with NATO, on August 26, and a later meeting in Moscow. On August 21, I met with Lehman in his office to assess his instructions. The NSC had given him four options; all were easily within the scope of my analyses, and I told Ron that all of them could be analytically justified as being in our national interests.

The many Geneva assignments and experiences had led to my continuing responsibility to draft instructions for the D&S negotiating group.

Round VI

On September 13, Avis Bohlen returned to Geneva as Executive Director for the NST delegation. Verne Wattawa, now a civilian, replaced Colonel Kramer as her principal administrator. Verne, later promoted to the Senior Executive Service, replaced Avis in that position in all subsequent rounds.

With Verne back on the delegation, the functional aspects of the work (such as a car, secretary and plenary participation) immediately became much easier. However, the workload for all of us in Round VI remained excessive. And now I had simultaneous, but separate, work to do on my own for all three ambassadors concerned with the START and D&S negotiating groups. Each ambassador required variants of the analyses I had prepared during the summer.

My work on the D&S Negotiating Group's instructions back in the US led to individual tasks now to draft three of the eight D&S plenary statements in Round VI. Each statement required five to eight versions before the ambassadors' final approvals. The plenaries also required preparation of several memcons describing our post-plenary negotiations. These were accomplished with Greg Suchan, Bob Barry and with Captain Ed Melanson, USN. The established routine for plenaries required at least three 15-hour days per week.

Ambassador Kvitsinsky, who had been responsible for the overall Soviet NST delegation and some aspects of the Soviet–Afghanistan War, was assigned as Soviet Ambassador to East Germany. He also took one of my interlocutors, Grinin, with him. During the first plenary of Round VI, Ambassador Obukhov, now head of the Soviet NST Delegation, showed his usual tough (read "intractable") stance.

My analytical work for the ambassadors was remanded to the September weekends. I completed the 50 and 30 percent force reduction studies for Ambassadors Lehman and Cooper, followed by a draft for Hank's arms control speech scheduled for presentation to the Strategic Air Command (SAC) in Omaha.

On October 14, we received a two-hour debriefing on the meetings between President Reagan and Soviet Premier Gorbachev in Reykjavik, Iceland. The heads of state had moved the negotiations farther in two days than we had during the last two years of the essential groundwork in Geneva. Their expansive discussions covered all aspects of INF and START, including the complete elimination of ballistic missiles. When the Soviets attempted to discuss elimination of his SDI Program, President Reagan surprised them by abruptly ending the negotiations a day earlier than expected.

Our negotiations improved after the Reykjavik meeting. Both delegations had extensive discussions with their Washington and Moscow superiors to work out new instructions for the remainder of Round VI. However, success was limited because both sides selected the parts of the Reykjavik meetings that supported our preferences and ignored the remainder, which, in the case of the US delegation, especially meant ignoring the part about total elimination of ballistic missiles.

The tours in Geneva were made tolerable through frequent parties given by the ambassadors, the Grassleys, and weekend dinners with Bob Barry and Verne Wattawa. The annual Marine Corps Ball was held at the International Hotel to celebrate the Corps's birthday on November 7. Customarily, the oldest Marine present, Ambassador Kampelman in this instance, has a *pro forma* dialogue with the youngest Marine on the Mission guard force. Max had been a conscientious objector during World War II who voluntarily starved himself to the conditions found in German concentration camps. This had the intended effect of enabling medical people to experiment in developing rehabilitation practices that would help to return the starved internees to normal health. He later became convinced that military force had a legitimate role in international relations and served as a captain in the Marines in the 1950s.

End of Round

The fourth through eighth D&S plenaries were complicated by the mixed signals we all received after the Reykjavik Summit Meeting, personal and professional preferences within the D&S Negotiating Group, and pressure from the Soviets for additional plenaries.

Intense higher-level activity continued to the end of Round VI. Ambassador Kampelman attended a ministerial-level meeting in Vienna and briefed us when he returned to Geneva on November 7. Soviet Ambassador Karpov asked Kampelman for a special joint plenary meeting at the US Mission at 4:00 p.m. on the same day. This was typical Soviet behavior for end-of-round initiatives.

The eighth and final D&S Plenary was at the US Mission on November 10. The Soviets hosted one last reception the next evening. We finalized all of the Round VI documentation on November 12 and then wearily flew home.

Meanwhile, Vic Alessi had been promoted, and Mike Mobbs had brought in Steve Maarannen, a man of hard-right persuasions, from the Los Alamos Laboratories, to be my new boss. The ACDA-appointed line management now were all affiliated and guided primarily by ties to OSD, specifically Richard Perle and Frank Gaffney. All this did not bode well. Mike Mobbs said that I was going to be removed from the delegation and stay in Washington to assist his new D&S Manager. I was also precluded from securing a long-desired transfer to the START Negotiating Group.

Bob Barry said that Mobbs was removing me as Senior Advisor to the delegation to make it easier for the OSD people to prevail, especially in the bubble meetings. Without my expertise, reputation and attitude, they could reassert their customary dominance on military issues.

With only a year remaining in my Schedule B Appointment and four levels of hostile management, plus OSD, my future in government service did not look very promising. Vic Alessi suggested going to Geneva as expected for Round VII and then trying for a transfer to the NSC or OMB in the second quarter of 1987.

Ambassador Kampelman was aware of my situation and had already talked with Director Adelman. We met on December 15. He also suggested trying for a transfer to the NSC. Separately, Bob Reisinger said that he would use his consulting firm to protect all my security clearances until a new job could be arranged.

Ambassadors Adelman and Cooper and Verne Wattawa were in a high-level meeting in Geneva in early December. Verne called me on the secure telephone on December 4 and I promptly circulated a summary of his report in ACDA and the State Department. Following Hank's arms control status briefing at SAC Headquarters, I received my first personal call in several years from SAC. Colonel J. Kelly asked for the testimony I had given to the House Armed Services Committee three years earlier.

The D&S IGs and WGs continued to support the delegation and I attended as the ACDA representative during December. On December 2, Kent Stansberry was purveying the OSD party line on offensive weapons during an IG meeting, and I challenged the OSD work because their spectrum of possibilities ranged so bradly as to be useless: 250 percent. I referred Kent to the study I had done in Geneva ten months previously, which was both more specific and useful.

Ambassador Kampelman, who was very sensitive to nuances in Soviet behavior, spoke to me on December 12, 1986. The Iran–Contra mess, which the public story attributed to over-zealous NSC staffers MacFarlane and Lt. Colonel North, USMC, had weakened the Reagan Administration's domestic and international credibility. Contrary to his expectations, the Soviets avoided exploiting the weakness; they simply did not mention it. Kampelman said that he wanted to determine the reason behind this anomaly in Soviet behavior. However, such an effort would have to await Round VII in Geneva because I would need various very sensitive Top Secret documents that Bob Barry could provide, but only then and there.

My primary task now was to draft the D&S Negotiating Group's instructions for Round VII, based on all of the recent developments in Reykjavik, Vienna and Geneva.

Chapter 23. NST Round VII & Space Policy

After a day of last-minute revisions to our instructions for Round VII, we departed from Dulles Airport on January 13, 1987. The NST Ambassadors flew to NATO Headquarters in Brussels on January 14 to brief them on the status and prospects in the arms negotiations. Ambassador Kampelman related a late-night call from Soviet Ambassador Vorontsov on January 15, canceling the opening joint plenary session and refusing a meeting of the Executive Secretaries.

Verne Wattawa was out for a week due to a death in the family, and Ambassador Cooper had me fill in on Verne's job as Acting Executive Secretary during his absence. There were also two special studies underway, one for Hank on nuclear effects and another for Max.

In Geneva, Bob Barry selected six or eight classified documents that would be helpful to my investigation for Max. These reports were primarily concerned with the USSR's current internal political, financial and military affairs.

Premier Gorbachev had more problems than solutions: (1) The entire apparatus was overburdened by the Soviet military establishment (publicly accounting for seven percent of their GDP, but more likely 20 percent); (2) MOM (the Soviet industrial community) was not fulfilling their unrealistic production quotas; (3) There were pressures from the Reagan Administration's strategic arms renewal and SDI programs; and (5) Soviet plans to alleviate these circumstances were unrealistic.

SDI was of particular concern to the Soviets because: (1) The US was usually ten years ahead of them in computer and sensor technologies such as those that would be needed to offset the new US initiatives; (2) They simply could not afford this new major round of strategic offensive and defensive arms; and (3) The Soviet–Afghanistan War was a drain and was not going well for them (thanks to $500 million in covert US aid to the Afghans, especially and recently the Stinger shoulder-fired, infrared-guided anti-aircraft missiles).

My principal conclusion, on the basis of the largely qualitative analysis and after correlating all of these factors with help from the CIA documents, was that the Soviets needed strategic arms reductions more than we did! Consequently, they had nothing to gain by exploiting the Iran–Contra affair.

On January 26, I handed the study results to Ambassador Kampelman. Bob Barry also reviewed the document and contributed to the final report. Bob later furnished this study to the intelligence community.

Ambassador Kampelman had frequent speaking engagements in the US and Europe. On February 13, he mentioned that he was using the unclassified aspects of the study in these speeches, suggesting that the Soviets had a greater need for arms reductions than the United States. The Soviets certainly did not miss this in their analysis of his public statements. The analysis was also furnished to Ambassadors Cooper, Glitman and Lehman, Verne Wattawa and Greg Suchan.

This study indicated, early in 1987, that the Soviet system of governance was very likely to fail. The actual event, about three years later, was sooner than expected. Further, the catastrophic failure was a relatively bloodless, internal affair. This outcome was also much better than I anticipated.

Negotiations

The D&S Group decided early in Round VII that there would be members-only participation in the US–Soviet plenaries. This meant that, although I was the senior advisor and quasi-member as ACDA's senior staffer on the delegation, Ambassador Cooper could keep me out. However, the advisors still maintained some influence. On January 26, in a D&S meeting in the bubble, I observed that of the 18 paragraphs in the draft plenary statement, 11 paragraphs exceeded our Round VII instructions. That gave pause to some of the more aggressive members.

Ambassadors Kampelman and Cooper asked for two more quantitative force structure variants of my studies. Eventually, the plenaries were re-assigned to the advisors with the usual blizzard of statement preparations, memcons and liaison.

The negotiations were more dynamic during Round VII. The ambassadors paired off privately and more frequently with their opposite numbers: Kampelman/Vorontsov, Cooper/Detinov, Lehman/Obukhov and Hamner/Zaytsev.

On Friday, February 27, the congressional delegation (Codel) visited Geneva, accompanied by Richard Perle from OSD. We held a reception for them and the Soviets at the US Mission. After participating in the last Joint Plenary at the Soviet Mission on March 2, we arrived home on March 8.

New Milieu

I had been in Geneva as ACDA Senior Advisor for four consecutive rounds of the negotiations. During the past three rounds Mike Mobbs and his associates had been attempting to have me removed. In the middle of Round VII, they sent USAF Lt. Colonel Charlie Phillips to join us in Geneva to gain some experience and prepare to be a replacement on the delegation.

On March 12, Mike Mobbs said that I would not be returning to Geneva for Round VIII, as the Defense and Space Division workload required me to stay at ACDA for the last nine months of my appointment. On March 21, Ambassador Kampelman had a heart attack. Fortunately, he recovered rapidly but was recuperating at home. Without the support of Ambassadors Kampelman and Cooper, my prospects of returning to Geneva were zero. Thus I went about the customary stateside ACDA tasks and polished my resume once more.

Back-Stopping

Due to a trip to Moscow by Ambassador Cooper and others (but without Ambassador Kampelman), Round VIII was substantially delayed. The Round VIII instructions were re-drafted to include results from the Moscow meeting and were delivered to the White House on April 28.

Lt. Colonel Phillips found being my replacement in Geneva to be quite a chore. He became very testy in his secure telephone reports and requests. In the middle of our routine dialogue on June 22, he simply hung up the telephone. Charlie returned to the US on June 26. Mike Herlihy was sent to the D&S Negotiating group as his replacement. They both had something of a shock as the job requirements became more apparent. At half of my workload and age, they both soon found other assignments.

Mike Mobbs finished his two-year ACDA career in late June, but Steve Maarannen remained. Round VIII lasted over 13 weeks. Characteristically, the Soviets tabled a major Defense and Space Treaty proposal with several protocols on July 29. This was intended to augment or replace the existing ABM Treaty, a continuing issue in Round VIII. The Soviet Krasnoyarsk ABM radar installation was also becoming a contentious subject in the negotiations. This huge phased-array radar was located in the south-central part of the USSR and sited with a clear mission of battle management rather than the permissible role of peripheral surveillance.

Space Policy

In mid-1987 President Reagan decided to review US space policy. He was motivated by effects of the shuttle disaster, the impact of the grounded space shuttle on our military and civil space programs, and the news that the NASA budget estimate for the Space Station had doubled.

A Presidential Initiative requires a more vigorous approach than the usual process of Defense and Space IGs and WGs. There would be a Core Group of representatives from the NSC, OSD, State, ACDA and NASA. On August 5, I became the reluctant ACDA representative for this National Space Policy Review.

The Core Group met frequently at the nearby New and Old Executive Office Buildings (NEOB & OEOB). The Office of Management and Budget (OMB) was also located in the NEOB. Their representatives frequently joined our meetings. We were also joined periodically by senior managers from our agencies and representatives from the NSC, CIA and DIA.

A lot of our effort in the Space Policy Review could have been eliminated if NASA had accepted the *Ride Report.* Sally Ride, an experienced and thoughtful astronaut, had headed an earlier NASA panel on future space activity. Her report emphasized near-space, usually unmanned earth-space science, for the more immediate and tangible benefit of mankind.

Getting to know the NASA senior officials and planners did not inspire confidence. I found them arrogant and uniformly pre-occupied with getting more money for pet projects. Their primary interest was developing large manned programs like the Shuttle and Space Station to give them big public relations advantages. NASA planned manned space activity for this reason rather than any fundamental justification for man's role in many space missions. Advancing truly scientific earth and near-space science and aeronautics was seldom on their agenda.

NASA maneuvering within the government had forced DOD to give up most of their military space launches to provide these payloads to the NASA Space Shuttle.

The total military, scientific and civilian payloads enhanced their rather flimsy justification for the manned Shuttle program. However, the Challenger disaster and two-year hiatus in shuttle launches was seriously jeopardizing the replenishment of aging, and essential, military and weather satellites.

The NASA representatives tried to ignore the problems created by their shuttle predispositions and performance. The Air Force eventually initiated a new unmanned space booster program to launch the most urgent payloads.

On October 19, I had a chance encounter with Sam Brow, an assistant to Vice President George H.W. Bush. He was aware of my space policy review. He said that the Administration was considering a management initiative to establish a permanent National Space Council. I encouraged him, since the current Defense and Space IG and WG process and our policy review were too slow and bureaucratically encumbered to provide timely support to the NST delegation in Geneva. The endless meetings were producing little new and useful policy guidance.

Next

My four-year appointment was to terminate on December 29, 1987. To be considered as continuous government employment, I needed to secure a new job that would begin within 48 hours — next to impossible. Presidential personnel did their best to help me find a new job, but were unsuccessful.

After six months of innumerable close encounters, and let downs, Verne Wattawa talked with Ambassador Kampelman about providing another year on the delegation, not in ACDA but as a State Department employee.

On December 18, Ambassador Cooper completed a position description form for a Special Assistant to the NST Defense and Space Negotiator. The State Department approved Hank's job description and a new Schedule B appointment. On December 24, Verne called to say that all of the paperwork had been cleared at State and that I should plan on returning to Geneva on January 9 or 10.

Now I had to get timely security clearance transfers. I spoke with Sharon, Ambassador Kampelman's personal secretary, on December 28. The Ambassador immediately provided a memorandum to Ray Firehock in ACDA Security, and the deal was done.

Chapter 24. NST Rounds IX & X

During my nine-month absence from the NST negotiations in Geneva, the milieu had become much more dynamic. While the normal routine of plenaries remained fundamental to the arms negotiations process, there were many important peripheral activities. These included the summit and ministerial meetings and more frequent one-on-one, two-on-two and three-on-three sessions between ambassadors and group negotiators.

The growing rapport at summit meetings between President Reagan and Premier Gorbachev went hand in hand with further progress in the arms negotiations. The new Soviet buzzwords of "glasnost" and "perestroika" and the President's motto of "Trust but verify" had become an important leitmotif to the negotiations.

Ambassador Cooper was the US Negotiator for Defense and Space, now assisted by William H. Courtney, also from the State Department. Our negotiator for strategic offensive arms was Ambassador Stephen R. Hamner. His Deputy Negotiator was Rear Admiral Dean R. Sackett, USN, from the Joint Chiefs of Staff. Ambassador Mike Glitman remained the effective leader of the Intermediate Nuclear Forces (INF) Group. Their negotiations were progressing very well and independently of the D&S and START Negotiating Groups.

Ambassador-at-Large A.A. Obukhov was still the USSR Head of delegation and Major General Yu. V. Lebedev was his Deputy. Ambassadors Kuznetsov and Masterkov headed their Space Arms and Strategic Arms Groups, respectively. The dignified and aging Lt. General N.N. Detinov remained as the Soviet delegation member of the group on space arms.

My role on the NST delegation had changed somewhat. In addition to normal duties as advisor, I was primarily Ambassador Cooper's Man Friday (and usually Saturday and Sunday, too). I also worked for Verne Wattawa by editing many of the memcons and cables that he sent to Washington, and occasionally substituted for him as Acting Executive Secretary when he was stateside or elsewhere in Europe with our ambassadors.

The Geneva negotiations had become much more publicized affairs. Ambassador Obukhov arrived in Geneva on January 12 with a press conference pronouncing the linkage between space defenses and strategic offensive arms reductions. Ambassador Kampelman responded on January 13 to an audience of about a hundred reporters and cameramen at the US Mission. The first Joint Plenary was held January 15.

Analyses

Hank wanted an update of my US and USSR force structure studies to reflect all of the developments during the nine-month absence. Once again, I summarized the US Triad strategic forces, reflecting the new US and Soviet arms reduction proposals and their latest position regarding a mutual reduction to 3300 strategic nuclear delivery vehicles (SNDVs), now defined as single weapons, and produced updated missile and bomber charts, followed by a time-phased strategic weapons chart. Another six charts were completed, some of which delineated the extra bomber weapons not covered by the mutually proposed treaties on SNDVs.

Success sometimes has two sides. Hank now wanted the same analysis for the Soviet strategic forces and views on which strategic weapons they would retain or retire in realizing a new and lower limitation of 3300 SNDVs. Hank got his report in time to take it to the States, where he had an agreeable discussion of the charts with Lt. General Scowcroft (Ret.) at the White House.

A concurrent analytical task was more immediately in support of the D&S negotiations. The US and Soviet sides differed on the definitions of "treaty" vis-à-vis "agreement." In the US view, a treaty was a formalized and Senate-approved pact. An agreement was a less formal but equally binding obligation. Hank wanted everything that had been said on the subject in the prior eight rounds and a summary for the purposes of negotiations. Eleven days and 35 pages later, he was given the report on Sunday, January 31.

Early in Round IX, the Soviets tabled a new START Treaty. Ambassador Cooper had me analyze the contents and determine the implications, particularly as to how their initiative might affect the D&S negotiations. This task was included in the overall strategic forces analyses described above.

On February 19, Hank and the other delegation ambassadors went to Helsinki and Moscow for another ministerial meeting. I updated my work on the Soviet proposals and maintaining strategic stability during mutual arms reductions.

The NST delegation had lacked the capability for independent analysis of strategic forces for most of 1987. My prior work had kept Ambassadors Kampelman, Cooper and Lehman well ahead of the recalcitrant efforts of DOD and JCS in providing analytical work for the strategic arms negotiations. While the delegation still had to wait for official Washington channels to function and provide positions and instructions, our ambassadors found my delegation work useful for early perspectives and a means to motivate better stateside military responsiveness.

Negotiations

In addition, my role in the D&S negotiations was still very similar to that of the previous five rounds in Geneva during 1985–1987; I was responsible for daily liaison with the START Negotiating Group, preparing plenary statements and memcons, frequently serving as Duty Officer for the NST delegation (the workload always required being at the Mission on weekends anyway), and participating in meetings of the US–Soviet Working Groups.

The past three years established mine as a familiar face to the Soviets and I like to think they recognized in me as someone interested in serious negotiations for mutually beneficial arms reduction — unlike several obfuscating and obstructive associates among our delegation.

Round VIII had produced D&S and START draft treaties from both the US and Soviet sides. Evolution, refinement and concurrence on these draft documents were now the focus of our negotiations. The sides developed a practice of side-by-side comparison and bracketing of the many differing positions. When common language and positions were established, these paragraphs became unbracketed portions of a potential overall agreement. One of my assigned tasks was to keep both draft treaties up to date for the D&S Negotiating Group.

Our instructions and plenaries were primarily directed to further US views and positions that would produce a favorable evolution of the draft treaties and protocols. The first Joint Plenary of Round IX was held at the Soviet Mission on January 15. We had a post-plenary bubble-room session with Terry Schroeder, our very likable public relations officer. We cleared a cable for his use, emphasizing the now more public motif of the arms negotiations.

On the third D&S Plenary at the US Mission on January 22, we tabled a new version of the US draft treaty. Consistent with the new public interface, Ambassador Cooper held a press conference afterward. Hank was prominent on the local German and Italian TV channels that night.

Colonel Dan Gallington persisted in monopolizing bubble-room discussions and reducing the D&S Group's effectiveness. As the Reagan Administration waned, some of Dan's DOD sponsors accepted more lucrative jobs in the private sector. Dan continued his delaying tactics. On April 13, he produced another plenary statement for consideration. With Hank out of town and Bill Courtney and Verne Wattawa otherwise occupied, as the ranking State representative I was the default chair. I rejected his paper without group discussion and asked him to re-do it.

Following three years of negotiations and several ministerial and summit meetings, near the end of Round IX, the US and USSR had identified their areas of agreement and remaining differences. ACDA provided an Issues Brief on May 6, 1988, paraphrased below.

There was general agreement to effect phased 50-percent reductions to equal levels of strategic offensive forces over a period of seven years, following ratification of the treaty.

The number of weapon delivery vehicles deployed would be limited to 1600 ICBMs, SLBMs and heavy bombers. Although all heavy bombers carried multiple weapons, each bomber was to be counted as only one delivery vehicle and one warhead. (This latter condition was very advantageous to the United States.)

The mutual ceiling would be 6,000 warheads. The sides had also agreed on a sublimit of 4900 ballistic missile warheads, but differed on the mix of ICBM and SLBM warheads. The aggregate throw-weight of ICBMs and SLBMs would be reduced to 50 percent of existing levels for the duration of the treaty.

The sides differed substantially regarding treaty verification by on-site inspections, banning mobile ICBMs, and the allowable ALCM range from heavy bombers. The most serious disagreement in the US–Soviet positions remained the contingency conditions related to Defense and Space. The US would not accept a contingent interrelation of START and D&S agreements. The Soviets would not accept 50 percent force reductions without an agreement on D&S issues. They still intended that any agreement would eliminate or greatly inhibit President Reagan's SDI Program.

With intensified arms negotiations and the US presidential primaries in full swing, the Senate Observer Group (Codel) had many Senators visiting the NST delegation in Geneva. The Democrats were now in control of the US Senate and outnumbered our Republican visitors. Senator Ted Stevens (R–Alaska), was noticeably less enthusiastic in his Minority position.

Very early one morning (or very late at night in Washington), while I was alone and making coffee, two visitors arrived who did not know the elevator codes. I greeted Senators Al Gore (D–TN) and Jay Rockefeller (D–WVA) in the Mission lobby, gave them coffee, and took them to the bubble for an impromptu briefing on the status of the NST negotiations.

On May 11, senior diplomats and military officials arrived from the US, including Ambassadors Kampelman, Lehman, Redman, Rowney and Nitze. The following day the US Secretary of State, George P. Shultz, arrived. Soviet Ambassador Karpov returned to rejoin his former delegation colleagues, Ambassadors Obukhov, Lebedev, Kusnetsov, Masterkov and General Detinov. Of these, Detinov (one of Hank's long-time counterparts in the negotiations) was perhaps the most approachable. We frequently met when he was walking his adored grand-daughter around the Soviet Mission.

Ambassador Glitman had succeeded in his four-year effort to forge a mutually agreed US–Soviet pact on Intermediate Nuclear Forces (INF) deployed in Europe and Asia. All of the visitors had arrived to finalize an INF Treaty that would eliminate these nuclear weapon systems.

The successful negotiations were characteristically intense. Karpov and Obukhov exhibited their usual negotiating tactics by adjournment at 2:30 a.m., followed by a sudden request to meet again at 5:30 a.m. This was to take advantage of Ambassador Kampelman's weariness and declining health. Verne Wattawa spent the entire night supporting Max in this hectic negotiating environment.

Several of us waited in vain for Verne to join us at a restaurant near the US Mission. We finally gave up, ate, and departed around 11:00 p.m. The next morning when I opened the office refrigerator that I shared with Verne, I saw that our cache of beer, cheese and bread was missing. All that remained was a note: "Thanks for the food, Colin." General Powell had made a late-night visit.

End of Round

May 23–25 were the last busy days of the very welcome end to Round IX. The usual D&S End-of-Round Report and various over-due memcons were drafted.

Due to the absence of the ambassadors I was to be the ranking diplomat for a May 24th Soviet D&S visit to Hank's residence. Since Barbara was visiting me in Geneva, this was her first experience of a social meeting with the Soviets.

At this juncture the Soviets were proposing a joint manned exploration of Mars. That was General Detinov's agenda for the evening. Referring to my several months on President Reagan's new Space Policy Review during my absence from Geneva, I offered a diplomatic demurral.

We both agreed that for the next decade or more, the joint US and USSR effort on the Space Station would likely suffice as our principal common space effort.

Back in the States, Hank gave me another analytical task regarding various sensors for START and D&S, with 29 questions and answers to be prepared for the NST negotiations and the Congress. The relevant contents from the recent Summit Meeting were also integrated into the task. Meanwhile, we were also drafting new instructions for Round X. The Soviet Krasnoyarsk ABM radar remained a serious point of contention for future negotiations.

Round X

We returned to Geneva on July 9, and on July 11, Ambassadors Kampelman, Cooper and Lehman held their first Round X press conference. ACDA was called to determine the status of our instructions that had not yet been received.

Hank provided two more long-term analytical tasks: (1) studying our current START instructions and attendant issues and (2) summarizing all of the dialogue in Round IV (1986) on the ABM Treaty. Hank had become increasingly involved and effective in both the US and Geneva discussions on all aspects of offensive and defensive arms control. He also hinted that we would travel back and forth to the States together during Round X.

Verne confirmed our probable travel plans the next day. We were to leave July 21 and return August 10. Besides everything else, I was keeper of the most up-to-date versions of the D&S and START draft treaties. These documents were to be carried in a diplomatic pouch on every trip — essentially I was serving as Hank's "bagman," library, and analytical support on these trips.

Our first Joint Plenary was followed by a D&S Plenary on July 19. Being the volunteer Duty Officer every weekend became routine; I was at the Mission anyway for other tasks, essentially working round the clock.

Stateside–1

We went back to the States as scheduled. Carrying a diplomatic pouch stuffed with about 20 pounds of secrets and a non-professional courier pass (there are people who do this for a living) always caused delays at foreign airport terminals.

Hank needed a quick update on the US and Soviet bomber force weapons, those covered by the proposed treaties and all others, as he was on his way to a conference in Colorado Springs. I also touched base with Ambassador Kampelman's office and offered to provide a personal update on the Geneva negotiations.

Bob Reisinger arranged another meeting at AFSC Headquarters. I provided them with my latest analyses, but they had little new data. I also had lunch with Bill Beecher, and later, Joe Jennette from IBM.

Consistent with his expanding influence in top-level defense issues, Hank had sent a memorandum to General Colin Powell from Geneva that was very critical of the new compromise defense budget structure and followed up when stateside. As a consequence, President Reagan vetoed the budget proposal on August 3.

On August 4, I briefed Ambassador Kampelman on the latest force structure analysis. He provided another analytical task to perform in Geneva. The next day Hank told me to examine the recent and formal ABM Treaty Review from a unilateral and bi-lateral perspective, including the protocols and relevant plenaries. I made an oral summary on August 8, and he asked me to give this report to Lucas Fischer in ACDA.

I kept in touch with Verne in Geneva to summarize everything that we had been doing in the US. Hank was so busy by now that I was reviewing his mail as well as the arms control cable traffic. With Hank again on the West Coast, I made a return trip to Geneva on August 13, using another courier pass.

Geneva

On August 19, Pakistan President Zia was killed in a plane crash along with the US Ambassador and an American General Officer. This event later proved to be an assassination. We were all worried about Greg Suchan because he usually traveled with these people, but Verne, Hank and Dan Gallington finally confirmed that Greg had somehow missed the fatal flight.

On August 22, we gave a farewell party for Colonel Dan Gallington, who was returning to the Pentagon after ten rounds in Geneva.

My work continued on summarizing Round IV for relevance to the ABM Treaty limitations on space defense research and development. Hank and Verne were briefed the preliminary results on August 22. During Hank's next visit to the US, I stayed behind in Geneva to work on the ABM Treaty analysis and cabled him the results on August 31.

General Burns (Ret.), now Director of ACDA, and Ambassador Kampelman returned to Geneva to meet with Ambassador Karpov on August 30. Their dialogue continued during a reception for the Soviet delegation at the US Mission the following evening. Klapovskiy, a long-time interlocutor of mine during the negotiations, had skipped Round IX. During this reception, we had an extensive exchange, making up for lost time.

SDI

During our prior visit to the US, Hank had arranged for Lowell Woodward, a scientist from the Lawrence Livermore Radiation Laboratory (LLRL), to give us a briefing on "Brilliant Pebbles." This was a space-based ABM system designed to collide with Soviet warheads in mid-course.

Brilliant Pebbles was to be a computerized, highly maneuverable, independent, space-based system which could sense, intercept and collide with incoming strategic missile re-entry vehicles. Several thousand of these small interceptors would be placed in permanent orbits consistent with the apogees of ballistic missile trajectories. When an attack was detected, the "Pebbles" could be selectively activated by a ground-based battle management system.

A group of 27 LLRL scientists, which included Drs. Woodward and Teller, had successfully developed and tested a prototype vehicle in their laboratory. Apparently, President Reagan's SDI Program had been successful!

Within their instructions and security constraints, diplomats have some latitude in presenting nuances that can favorably affect negotiations. Given Dr. Woodward's success in Brilliant Pebbles, which at the time was very closely held, I took a new tack with Klapovskiy. No mention was made of Brilliant Pebbles but for eight years the Soviets had been tracking my technical and analytical work with the US Congress and in Geneva, and they knew of my apparent working relationships with Ambassadors Kampelman and Cooper. My exchange with Klapovskiy was intended to further motivate the Soviets to achieve an agreement.

Acting Executive Secretary

Whenever Verne was otherwise engaged, attending NATO or Ministerial meetings, he asked me to fill in for him by clearing all of the D&S cable traffic to Washington and dealing with the Soviets to set up meeting arrangements with G.A. Zaytsev, Executive Secretary, and M.N. Lysenko, his Deputy for the Group on Space Arms, at the Soviet Mission.

Stateside–2

On September 13, Ambassador Cooper returned to South Carolina but told me to meet him in Washington on September 20. I reviewed Hank's US and Geneva cables and his mail to brief him on arrival. Hank was in Washington for another Ministerial Meeting. I gathered talking points from Ambassador-at-Large Nitze's office and discussed with Hank how these factors might affect his latest D&S Treaty Proposal.

On September 27–29, Hank also participated in the Defense Science Board Task Force. He had more irons in the fire than a blacksmith. I provided nightly summaries of his cables, mail and progress in the Geneva negotiations.

Geneva Again & Again

I returned to Geneva on October 1. A D&S Plenary statement was prepared for October 5. The Washington scene was de-briefed to the D&S Group and Hank later gave us a higher-level update. He was very tired and pale. The endless negotiations of 1988 were beginning to wear on all of us.

This was very likely my last year in Geneva. In some positions, older higher-level workers are cushioned by assistants, discrimination and self-determination. However, my current position did not provide that latitude. IBM used to say that if a man retired at 60, he lived an average of 13 years; if he retired at 65, his life expectancy was an average of 18 months.

Bob Barry and I had been entertaining Klapovskiy and other people on the Soviet delegation at local restaurants for the past four years. The Soviets sponsored few such meetings below the ambassadorial level, but Klapovskiy finally invited us to a first-ever lunch at the Soviet Mission on October 14. On their home turf, Klapovskiy and Pokhomov were congenial hosts and more responsive than usual. From Klapovskiy's comments, we deduced that this would be his final round in the negotiations.

Ambassador Kampelman cabled to ask me to contribute to an article for *Foreign Affairs* related to strategic stability. Simultaneously, Hank needed a special paper drafted for his Defense Science Board meeting in the US on October 13–14. He also asked me to review the outline for his intended Anti-Satellite (ASAT) paper. Both of Hank's papers were refined several times during the next two months.

Stateside–3

Hank Cooper had another series of meetings at the DSB during October 26–28. With his secretary, Jean Swoyer, we went back to the US again on October 25. My usual stateside tasks went on: handling Hank's communications and long, secure telephone calls with Verne to keep everyone informed regarding Washington and Geneva activities.

One errand required meeting with the D&S IG and WG, and separately with the DIA and CIA to develop a few paragraphs of new instructions. I took these to the White House (OEOB) offices of the NSC for final approval.

On October 28, Hank actually told me not to work the weekend, nor to take a diplomatic pouch back to Geneva. At long last, negotiations were beginning to wind down on both sides of the Atlantic Ocean. The following week was mostly taken up with errands with ACDA Legal on definitions and locating a National Security Directive (SNDD-318) for Hank.

November 5 was my last flight to Geneva. Ten trips across the Atlantic in 1988 meant about 60 days of jet-lag had to be worked off. It was easy to understand why some airline pilots prefer north-south routes in their later years.

On November 8, George H.W. Bush was elected to be the 41st President of the United States. This choice was more favorable for arms control than Dukakis might have been.

During Verne's absence, I scheduled several D&S Group meetings with S.V. Kulikov and V.F. Saratov, now Executive Secretaries for the Soviet delegation. This included the usual routine of plenaries, a two-on-two meeting, and a luncheon for Bill Courtney and General Detinov. The twelfth draft of our last D&S Plenary Statement for Round X was also reviewed. That evening Hank and Bobbye Cooper gave the last of many enjoyable receptions for the D&S Group at their residence.

November 9–18 saw the usual hectic end-of-round activities. The last D&S Plenary was at the US Mission on November 9. Our reception for the Soviets was held on November 14. Ambassador Kampelman attended the final Joint Plenary of Round X on November 15. The advisors worked on the cables to the US and the End-of-Round Report while the NST delegation leadership made their customary trip to NATO in Brussels.

I arrived back at Dulles Airport for the last time on Saturday, November 19. After two days of recovery, I drove to Florida with Barbara for two weeks of R & R.

At the current rate of progress in Defense & Space and START, another strategic arms agreement would require several years. The immediate future of arms control negotiations was becoming more uncertain. Internal problems in the Soviet Union were becoming apparent. Ambassador Kampelman agreed that the State Department needed the kind of analytical capability that I had provided to the NST delegation ambassadors in Geneva. With temporary extensions of the present job, they kept me on while I looked for another, less demanding, position at ACDA, State or the NSC.

Chapter 25. One Last Strategic Study

Ambassador Kampelman informed me on December 23, 1988 that State's Political and Military Affairs Division (PM) was willing to have me continue my analytical work stateside in support of the Geneva negotiations, and Hank secured a three-month appointment extension to await developments. On March 13, the State Department extended my appointment to January 30, 1990.

Delays, confusion, uncertainties and personnel changes are characteristic of a presidential transition. This deferred the Round XI negotiations until mid-July, followed by Round XII in late 1989. Ron Lehman later became ACDA Director. Ambassadors Cooper and Hamner soldiered on for both sessions.

The increasingly apparent and progressive collapse of the Soviet Union in mid-1990 ended formal negotiations by the US and the USSR. Although my 1986 study for Ambassador Kampelman had anticipated the systemic failure of the Soviet Union, the event occurred earlier than expected.

Ambassador Kampelman recognized that they needed independent strategic forces analytical capability within the State Department, which should be provided by PM, the Political and Military Affairs Division. He arranged interviews for me with Jim Holmes and other managers in PM. But there was a natural bureaucratic reluctance to develop a new capability similar to that possessed by DOD and the NSC, with the potential for competition and conflict with these powerful organizations.

Meanwhile, the NSC was being reconfigured, and there was a possibility of going there. However, my prior experiences had shown that any new government job would take months to materialize.

I decided to do some interim analytical work in order to stay current. If the initiatives in the NSC or PM were productive, I would need a new intellectual basis for such work, a fundamental re-examination of premises, facts, and shifting trends. I undertook a very comprehensive study of the US–Soviet strategic balance during 1989. Intelligence projections were delineated into three progressive phases of mutual disarmament during the next 20 years. The central aspects of this study could be performed in three months, given my preparations and current database. This would be useful to many government agencies that were (very) slowly developing new policies and actions for the Bush-41 Administration.

I no longer had the very up-to-date intelligence support that Bob Barry had provided in Geneva; I needed current, authoritative intelligence on the expected evolution of the Soviet strategic forces. Through an intermediary, INR, the State Department's intelligence organization that served both State and ACDA, provided Volume III of the latest National Intelligence Estimate (NIE).

On February 1, I began my last major force structure study covering 1989–2019 in three successive ten-year phases of US and Soviet arms reductions. I covered Soviet and US force projections in two months. The first ten-year phase of the new study began with the current status of the NST negotiations and the arms reductions expected during 1990–2000. As in the arms negotiations, the study objectives were mutual reductions to eliminate the silo-based, destabilizing, counterforce ICBMs during three successive phases; to reduce the SSBN–SLBM forces to a deterrence-only mission; and to reduce the land-based ICBM force to a few hundred small, mobile and relatively invulnerable missiles similarly dedicated to the deterrence mission. There would also be comparable reductions in the US and Soviet bomber forces.

As a preliminary to full, annual counterforce exchange studies to ensure strategic stability during 30 years of mutual disarmament, I developed phased opposing strategic target structures for the US and USSR, with the aim of producing a policy planning document of enduring utility.

The new study identified only one offensive weapon system development needed to improve the effectiveness of the progressively smaller US offensive forces. The need and characteristics were defined for an earth-penetrating nuclear warhead to counter the very hard and often deeply buried strategic targets in the Soviet Union. The MIRVed re-entry vehicle would be a more powerful version of the earth penetrating warhead developed for the Pershing II IRBM. This warhead would be selectively deployed on some of our ICBMs and SLBMs. There has been occasional public mention of such a development for the past 18 years, but I would not likely learn if DOD and DOE ever undertook such a program.

Next, I distributed the study to permanent government officials at the higher levels in the Executive Branch, as well as close associates, established defense intellectuals, and the intelligence community.

Jim Woolsey gave the study to General Scowcroft and Arnie Cantor on the NSC. I also gave it to Bob Howard, head of the President's Office of Management and Budget (OMB), whom I had met during the 1987 meetings in the NEOB on President Reagan's space policy. I later learned that OMB had tried to support some of the ideas, but they were not being taken up by the new Bush-41 Administration.

Given the many recent years of unrelenting enmity from DOD, there was only one person in the Pentagon I thought might be interested in the paper. Andy Marshall was still the DOD Director of Net Assessment. Andy passed it on to his Strategic Systems Group.

As always, my best conduit to the intelligence community was through Bob Barry. I gave Bob a summary of the study on April 6 and a week later he took five copies of the paper, for "You never know who" in the DIA and CIA.

Grateful as always for the INR's assistance in refreshing my database, I took copies of the paper to Ray Firehock and Mark Lowenthal. INR later said that the work had stirred considerable interest and other recipients had called INR to confirm the validity of the intelligence projections in the study.

The 30-year paper was distributed first to my closer associates on the delegation to ensure that my solitary effort had not produced any major errors in fact or judgment. Verne Wattawa commented that too many bombers had been included in the force structures and, with his considerable experience, questioned some of the targeting priorities. Bill Courtney, a dedicated, traditional, foreign service officer (and later, ambassador) expressed concern — not with the content but with my entire initiative of writing and distributing the 30-year study.

Since this was the last such effort I would be making, I had no concern about organizational constraints and protocol. No one else seemed to mind, either.

The 30-Year Study was more widely distributed in mid-April. On April 13, a copy was given to Richard Burt's assistant. I met with Ambassador Burt on May 1, and mentioned that the study could be usefully expanded by a stability and verification analysis if he or State–PM so desired. And if not, then I expected to retire.

By June 1, aides to Richard Burt, Jim Timbe and Reginald Bartholomew had called. All of the senior people had decided that, in view of the increasingly fluid situation in the Soviet Union, the stability analysis I suggested would be premature; only Ambassador Nitze favored such an analysis.

Apparently, there was no further reason for me to remain in the State Department. There was a flurry of security de-briefings, exit interviews, and the sorting or shredding of classified documents. On June 2, 1989, several bottles of champagne were drained with colleagues at the S/DEL suite, signaling the end of over 40 years of service to the nation.

Chapter 26. Strategic Nuclear Parity

Today's Balance of Power

By 1981, the Soviet Union had established the capability to destroy all of the hardened military targets and industrial/urban targets in the United States in less than one hour. Maintaining nuclear parity — America's comparable capability to destroy most of the USSR (and in later years, Russia) — along with a rational outlook on both sides — remains essential to national defense. Russian strategic military capabilities are still the most serious, albeit latent, military threat to our national entity.

The data reported and discussed below are contained in The Military Balance - 2007 which is published in London by the International Institute of Strategic Studies (IISS). For many years this organization has been the most authoritative and reliable source of unclassified information on global military forces.

The Russian ICBM forces now consist of 86 SS-18s, Models 4 & 5, deployed in four fields. Ten MIRVs on each missile provide 860 nuclear weapons effective against hardened counterforce targets such as ICBM silos and launch control centers. They also have 100 SS-19s that provide another 600 counterforce MIRVs for a total of 1460 nuclear warheads.

Russia also continues to emphasize mobile ICBMs to ensure survivability for trans-attack launch or later retaliation with 270 SS-25s and 50 newer SS-27 land-mobile ICBMs. They also tested a new Topol-M mobile ICBM during 2006–2007.

The 506 Russian ICBMs are a welcome two-thirds reduction from about 1,500 deployed ICBMs during the Cold War. They have pragmatically negotiated down to this strategically adequate and economically more sustainable force.

The Russians have reduced their ballistic missile submarines to twelve Delta III, IV & V SSBNs and three Typhoons. This provides an active force of 232 SS-N-18, SS-N-23 and SS-N-20 SLBMs. These missiles are armed with over 2200 MIRVed nuclear warheads. A new Bulava (SS-N-30) SLBM was successfully launched in September and December 2005. The missile is expected to be deployed on two new

Typhoon submarines by 2009. This will add another 40 active SLBMs and another 400 MIRVs.

The referenced IISS publication lists US ICBMs; 50 MX (Peacekeeper) and 500 Minuteman III ICBMs that provide a total force of 2000 MIRVs. Five hundred Minuteman I & II ICBMs have been eliminated.

The US has 14 SSBNs in the active naval forces. Four of these Ohio-class submarines are armed with a total of 96 Trident C-4 SLBMs. The other ten of our SSBNs are armed with 240 Trident D-5 missiles. More conversions to the D-5 SLBM are underway. This total force provides over 2,600 deliverable MIRVs. Our lack of land-based mobile ICBMs places singular dependency on our SSBN/SLBM fleet to provide nearly invulnerable counterforce and countervalue missile forces.

The very significant strategic bomber forces deployed by the US and Russia are not prompt counterforce weapons. Their longer intercontinental delivery times (hours versus minutes, for missiles) remand the bomber forces to second-strike missions and survivable reserves.

The PRC is many years away from any strategic missile forces comparable to those of the US and Russia. China is believed to possess 46 ICBMs and one SSBN armed with twelve SLBMs. They are expected to soon add more SSBNs to improve and expand their deterrence-only strategic forces.

Overall, the US strategic military posture has improved in the 20 years since I was actively involved in these matters. The MX ICBM has substantially added to our prompt counterforce capabilities relative to those of Russia. When they agreed to reduce their ICBM forces by two thirds and the US cut our Minuteman deployments in half, both sides achieved effective counterforce parity. Gradual and further reduction of the remaining destabilizing silo-based ICBMs is a very worthwhile goal.

In the 1980s, the Soviets assigned two re-entry vehicles (MIRVs) against each hardened US missile silo. The initial attack would be an air-burst above the target to minimize airborne debris. Moments later, the second MIRV would be a ground-burst for greater earth shock and to ensure a probable target destruction of well over 90 percent.

Our ICBM fields contain about 610 hardened targets. Since the Russians can utilize back-up targeting to compensate for obvious launch failures, weapon system non-availability and unreliability, this adds about ten percent to the attacking force of over 1,200 MIRVs. The total of 1,350 MIRVs is well within the 1,460 MIRVs that arm the current Russian SS-18s and SS-19s.

Launch under confirmed attack (LUCA) remains a necessary response doctrine for the US ICBM and SLBM forces. The 213 counterforce ICBMs silos and launch control centers for the SS-18s and SS-19s can readily be destroyed by the 500 MIRVs mounted on our MX ICBMs. The 1,500 Minuteman III MIRVs can be used for back-up and less-hardened counterforce and countervalue targets. Our SLBM forces would share this role with some held in reserve due to their enduring survivability.

I believe that the capability of Mutual Assured Destruction (MAD) remains an important deterrent to any Russian initiation of a general nuclear war. They still possess the forces capable of destroying the United States in less than one hour; but our retaliation would also completely destroy their nation.

We can expect intervals of increased tension with both Russia and the PRC. For sixty years, the competence and diligence of US strategic military forces have been essential to the defensive strength of our nation. They will need our continued support and periodic modernization.

Milieu of the US Armed Services

After 60 years of admiring and working with the Air Force, I am familiar with their characteristics. This branch of the service is largely managed by pilots. By necessity pilots are smart, athletic and courageous men and women with an institutional history of nearly 100 years of successful military aviation. They have certain preferences that were and are hard to overcome.

The US Air Force was slow to move ahead with the earliest deployments of Atlas and Titan ICBMs. What pilot would not prefer a fighting chance rather than an underground bunker — and certain death — during the first hour of a general nuclear war? However dangerous their missions, they usually had decent food and a clean place to sleep when they survived.

When my advocacy efforts with Congress and the President's Commission on Strategic Forces helped to ensure that the Air Force could develop a small mobile ICBM (Midgetman), the Air Force and the President's Commission insisted on the hardening of the Midgetman Transporter-Erector-Launcher (TEL). This allowed smaller deployment areas near established bases and lessened the need for random dispersal over larger areas — with the attendant additional expense and discomforts of extended field exercises and deployments. However, such hardening was intrinsically unnecessary, added years to development, and was very expensive.

The Midgetman ICBM Program finally faded away without any missiles ever being deployed. The Russians, as of 2007, reportedly still maintain a relatively invulnerable force of 320 SS-25 and SS-27 single-warhead mobile ICBMs. This will be an important advantage if our nations agree to further mutual reductions or elimination of destabilizing silo-based ICBMs.

The US Army is primarily a fielded force in wartime. They expertly and diligently deployed the Pershing II IRBM throughout Europe during the Reagan Administration, assuring the weapon system's relative invulnerability. Given this expertise, I always privately believed that the Midgetman Program should have been assigned to the Army for development, production and deployment.

With Midgetman, the Army would have become the third branch of our military services to have a major nuclear weapon system capability for strategic deterrence. They would have supported and nurtured this new role and mission, just as the US Navy does with their submarine-launched ballistic and cruise missile programs. The targets assigned to the mobile ICBMs could have been similarly controlled by SAC Headquarters.

This preference to assign the Midgetman to the US Army was logical but I never mentioned it in my advocacy efforts. The Midgetman had enough obstacles to overcome without creating an inter-service firestorm over roles and missions.

National Problems

Every generation faces major national and international problems whose solutions are elusive and require long-term efforts to understand and address. While the foregoing chapters give primacy to maintaining nuclear parity, a wealth of other important issues remain to inspire the constructive engagement of all thoughtful persons.

In two difficult centuries, the United States became the most powerful nation on earth. If we are to continue to enjoy the position attained for us by previous generations, and improve our nation and society, it will require many broad, concerted and selfless efforts. Seek your own truths in these matters and strive to make us even better.

As President Lincoln stated in his Second Inaugural Address, "America is the last great hope of the world." The circumstances have greatly changed in the past 140 years, but not that basic truth.

ACKNOWLEDGEMENTS

Several individuals contributed to and supported my work throughout the decades; I think of them as my "white knights." Some of the most outstanding are listed below in appreciation of their intellects and willingness to risk their careers, at times, in our common causes. Through our work, we influenced the defense decisions of four US presidents: Eisenhower, Nixon, Ford and Reagan.

Edward M. Rosenfeld, Architect, US Army Office of the Chief of Engineers, re: Nike Program, 1953–1954

Colonel J.J. Henderson, USAF re: B-70 Program; 1958–1959

B-1 Program, Invitational Orders I & II:

Jack B. Trenholm, GS-16, Civilian Chief, B-1 SPO 1963–1983

Major General Douglas T. Nelson, B-1 SPO Chief, et al., 1963–1983

Dr. Antonio Cacciopo, ATIC Director; 1976–1983

Major General George Keegan, USAF Headquarters, 1976–1977

B-1 Program, MX ICBM and Soviet Threat; I/O-3:

Lt. Colonel Bob Reisinger, 1979–1989

Lt. Colonel Hal Gale, 1979–1989

Colonel Richard Wargowsky, 1979–1983

Brigadier General Delbert A. Jacobs, 1979–1981

General David C. Jones; Chairman of the Joint Chiefs of Staff; 1959–1961 re: B-70; B-1, MX & Threat 1976 & 1981

Congress 1981–1983:

John J. Ford, Staff Director, House Armed Services Committee

Roy Jones, Staff Director House Interior Committee

Many staffers for the House Interior and Armed Services Committees, and the Senate Armed Services Committees

Senator Garn, Senator Laxalt, Senator Stevens, Senator Warner

Congressman Cheney, Congressman Clausen, Congressman Daniel, Congressman Dickinson, Congressman Lagomarsino, Congressman Lujan, Congressman Nelligan, Congressman Price, Congressman Sieberling, Congressman Udall, Congressman White and Congressman Young

National Security Council:

Ron Mann, 1981–1983

R. James Woolsey, 1983 & 1989